THE FORM AND PROPERTIES
OF CRYSTALS

THE FORM AND PROPERTIES
OF CRYSTALS

An Introduction to the Study of Minerals and
the Use of the Petrological Microscope

by

A. B. DALE, M.A.

Fellow and Tutor of Newnham
College, Cambridge

CAMBRIDGE

AT THE UNIVERSITY PRESS

1932

CAMBRIDGE
UNIVERSITY PRESS

University Printing House, Cambridge CB2 8BS, United Kingdom

Cambridge University Press is part of the University of Cambridge.

It furthers the University's mission by disseminating knowledge in the pursuit of
education, learning and research at the highest international levels of excellence.

www.cambridge.org
Information on this title: www.cambridge.org/9781107456099

First published 1932
First paperback edition 2014

A catalogue record for this publication is available from the British Library

ISBN 978-1-107-45609-9 Paperback

CONTENTS

CONTENTS

PREFACE

In helping students to master the principles underlying the examination, measurement and identification of minerals, much difficulty has been experienced in finding a text-book simple enough for those with only an elementary knowledge of physics and mathematics, comprehensive enough to give them most of what they need in a volume of moderate size, and at the same time inexpensive enough to suit their purses. This text-book represents an effort to supply their needs.

No attempt has been made to include more than the outlines of the subject, but references for wider reading are given in the text and a bibliography is appended. The fundamental subject of crystal structure is treated in the scantiest fashion, since the experimental work upon which this depends is at present beyond the scope of the geological student; it is hoped that enough is given, however, to show that internal structure is responsible for most of the properties of the rock-forming minerals which the petrologist must study. Since the polarising microscope is the most important instrument in the petrologist's equipment, especial attention is paid to both the theoretical and the practical aspects of polarised light and its refraction through crystal media.

Although the book is designed primarily for students of petrology, it is hoped that it may also be found useful to students of chemistry and others to whom crystals are becoming more and more matters of interest and importance.

In conclusion, the author acknowledges her indebtedness to Dr C. E. Tilley, Professor of Petrology and Mineralogy in the University of Cambridge, for his advice and encouragement, to Miss D. I. Reynolds of University College, London, for her valuable help with the diagrams, and to many other friends who have given welcome assistance in various ways.

A. B. D.

NEWNHAM COLLEGE, CAMBRIDGE
October, 1932

CHAPTER I

THE NATURE OF CRYSTALS

INTRODUCTION

The thin, superficial layer of the Earth, the only part of this planet from which we have been able to obtain samples, is composed almost entirely of mineral matter, and for this reason, if for no other, the nature of a mineral would be a subject deserving considerable study. About 1000 different minerals have been identified, analysed, described and named, often unfortunately with a clumsy style of nomenclature; a minority only of these are of common occurrence or of much importance either from the practical or theoretical point of view, and perhaps no more than 100 are sufficiently distinguished to claim the acquaintance of the ordinary geologist.

Each mineral is characterised by definite physical properties such as hardness, lustre, specific gravity, and by a definite chemical composition. It is not always possible to write the chemical formula of a mineral in a form as exact as that of calcite, $CaCO_3$, or quartz, SiO_2; many mineral formulae have to be written after the fashion of that for garnet $(Ca, Mg, Mn, Fe'')_3(Al, Fe''', Cr)_2(SiO_4)_3$, signifying that although the general chemical constitution is of a definite type, the relative amounts of calcium, magnesium, manganese and ferrous iron may vary in different specimens, their total amount per molecule being chemically equivalent to three atoms of any one of them, while the relative amounts of aluminium, ferric iron and chromium vary in the same way. In some cases mineral families exist whose

DC

I

members can be arranged in a series according to the relative amounts of certain constituents; thus the plagioclase felspars, a family very familiar to the geologist, can be placed in a series whose end members have the definite formulae $NaAlSi_3O_8$, albite, and $CaAl_2(SiO_4)_2$, anorthite, while all the intermediate members consist of these two constituents combined in varying proportions, the physical properties of the member varying in accordance with its chemical composition.

As additional complications in the chemical constitution of a mineral, small amounts of impurities may be present, often modifying its natural colour, and in some cases a close examination reveals an apparently homogeneous mineral substance as a mixture of two quite distinct minerals.

It will be obvious that the determination of a mineral by chemical means is necessarily a somewhat long and involved process, and a readier method of identification, of the better known minerals at any rate, is by the study of their physical properties and their external form or morphology. Among the physical properties the field geologist relies mainly upon *colour, lustre, hardness* and *tendency to cleave*; *optical properties* are also of great importance and can be investigated fairly easily in the laboratory. Finally, a study of the *crystal form* that is assumed by almost all minerals is of very great assistance in their determination, and is vital to an understanding of their real nature.

MORPHOLOGICAL PROPERTIES OF CRYSTALS

The natural external shape of a mineral is not only an indication of its identity, but it is the visible expression of its internal structure. The shape of a mineral specimen may, however, have been determined to some extent by outside forces, such as restriction of space during growth, in which case the external form may be but little guide to the real nature of the mineral. Left to develop freely, however, minerals assume characteristic shapes, and an important part of the work of the crystallographer is the study of the laws governing these shapes.

Crystals vary greatly as regards the relative size and shape of their various faces, and each crystal tends to assume a characteristic form or *habit*. A crystal may be described as showing a flat or tabular habit, a prismatic or columnar habit, a needle-shaped or acicular habit, and so forth. Different specimens of the same mineral, if formed under different conditions, may exhibit different habits.

General Properties. The essential characteristics of a crystal result from the fact, now well established, that the atoms composing the crystal have grouped themselves in an orderly fashion, adopting a system that is the same throughout any one crystal and for all crystals of any one mineral*. In non-crystalline, or amorphous, substances, on the other hand, there is no definite arrangement; the atoms have come together in a heterogeneous mass like apples of second-rate quality that the fruiterer pitches higgledy-piggledy into a crate. Continuing the simile, the choice Ribstons, piled up in a pyramid, each apple resting

* A few minerals are found which crystallise in two or more distinct forms; the form adopted generally depends upon the conditions of temperature and pressure at the time of crystallisation. Thus silica crystallising at high temperatures takes one of the two forms cristobalite or tridymite, while at lower temperatures it crystallises in the commoner form of quartz. This phenomenon is known as *polymorphism*.

in the hollow made by three or four below it, provides an illustration of the arrangement of atoms in a crystalline substance.

Some minerals appear to be amorphous in structure; they assume rounded, irregular forms, either simple or more commonly consisting of a mass of irregular hemispherical lumps, a form described as *mamillated*, or irregular spheres, the so-called *botryoidal* form. A mineral of this appearance is not necessarily non-crystalline, however, for each lump may consist of a regular or irregular cluster of crystals, very small or even microscopic in size. Opal and chalcedony, both forms of silica, appear in mamillated masses; under the microscope chalcedony is shown to be made up entirely of crystalline fibres, but opal is still usually regarded as a non-crystalline mineral. Modern methods of investigation, however, are revealing some degree of crystalline structure in very many solids previously considered amorphous, and it seems probable that no solid is entirely lacking in orderly atomic arrangement.

As a result of regular internal structure a crystal may possess certain physical properties which show variation in different directions. A substance such as tissue paper, which is made up of fibres lying roughly parallel to each other, tears in a direction parallel to the length of the fibres far more easily and regularly than in the transverse direction. In the same way, in accordance with their internal structure some crystals can be broken along certain directions more easily than along others. The hardness, elasticity, optical, thermal and electrical properties of a crystal may all vary in this way with varying direction and so are known as *directional* or *vectorial properties*. Other properties, on the other hand, such as specific gravity, are obviously independent of direction.

A mineral specimen in the amorphous state, or one in which minute crystals are jumbled together without orderly arrangement, exhibits no vectorial properties, since its properties in any direction represent the mean values of the vectorial properties of the constituent crystals. The sharpness of a pin varies with the direction in which the pin is approached: a muddled heap of pins is equally sharp on all sides. In the same way the hardness, for example, of a specimen consisting of an irregular cluster of crystals will be equal in all directions and will be the average value of the hardness of any one crystal measured in various directions.

Another result of systematic internal structure is seen in external regularity of shape. If a crystalline solid is allowed to develop from a molten or dissolved state under favourable conditions it exhibits plane faces set at regular angles one with another and meeting in sets of parallel straight edges, that is to say, the developing crystal takes on a regular geometrical form which is noticeable at whatever stage the development may be arrested. Two different specimens of a mineral generally exhibit similar faces, and the angle between any two corresponding faces is found to be constant for every specimen examined. This *Law of the Constancy of Crystal Angles* is fundamental to the science of crystallography.

A crystal usually shows two, three, four, six, eight or even a higher number of faces which are obviously of the same type, while certain other faces just as obviously belong to another set of a different type. Each of these sets of similar faces is known as a *form*; thus a quartz crystal shows one form which consists of six long faces, m_1, m_2, ..., m_6, together making a hexagonal pillar or prism (fig. 1), each placed at an angle of 60° with its neighbours, while other faces occur in forms of three which make pyramids at one or both ends of the pillar. Faces of one form may be distinguished from those of another by shape in some cases, but this is an uncertain criterion since the shape of a face depends upon regularity of growth and the relative sizes of adjoining faces. Faces of any one form can always be identified, however, by their relative positions in the crystal and by the angles which they make with each other or with other faces.

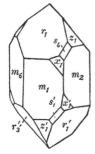

Fig. 1. Quartz crystal.

Moreover, by reason of the regular geometrical shape of a crystal, its faces are found to occur in *zones*—that is to say, in sets which meet in parallel edges (or which would meet in parallel edges if they could be sufficiently extended), and which form zones or belts running round the crystal. In fig. 1 faces m_1, m_2, ..., m_6 form one zone, faces m_1, x_1, s_1, z_1, ... form another, and faces m_1, s_1', r_1', ... yet a third. If an axis is taken through the centre of the crystal, parallel to the edges of any zone, and if the crystal is rotated about this axis, which is called the *zone axis*, the faces belonging to the zone will be turned successively into parallel positions. This can easily be seen by pivoting a quartz crystal between a finger and thumb so that it can be rotated about the long axis lying parallel to the edges of the hexagonal prism faces. If one face is made to reflect light from a window, each of the other faces gives the same reflection as it passes on rotation into the position originally occupied by the first face. Since the values of interfacial angles are characteristic of a crystal and lead to its identification, methods of measuring these angles are of great importance. The measurement is made by means of a *goniometer*. There are two main types of instrument, the more primitive *contact goniometer*, and the more accurate *reflection goniometer*, the acting principle of which is the same as that of the rough and ready demonstration described in the last paragraph.

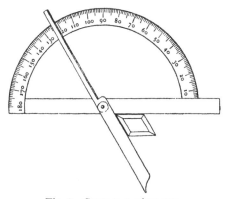

Fig. 2. Contact goniometer.

The contact goniometer (fig. 2) in its simplest form consists of a protractor to which is pivoted a rotating arm. The angle to be measured is fitted into the angle between the protractor and the arm, the crystal edge being perpendicular to the plane of the protractor. The size of the angle is indicated by the reading of the upper end of the arm on the scale of the protractor. The actual angle measured is almost always more than 90°, but it is customary to describe an interfacial angle by its supplement; thus the interfacial angles of the hexagonal quartz prism are described as 60° and not 120°.

The reflection goniometer (fig. 3), invented by Wollaston in 1809, is a more sensitive instrument and better adapted for work with small crystals. The crystal C is held by means of wax on the end of the horizontal axis of a vertical graduated circle which rotates with the axis, and the crystal is adjusted so that this axis is parallel to one set of

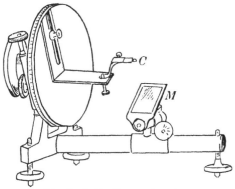

Fig. 3. Reflection goniometer.

zonal edges. A small bright object, such as a diamond-shaped hole cut in a window shutter, is reflected from one face of the zone to the eye of the observer, and the crystal is rotated about the axis until this image becomes coincident with an image of the same object reflected by an inclined mirror M set vertically below the crystal and parallel to the axis. If the crystal is accurately adjusted, as the axis is rotated an image from each face of the zone in turn will become coincident with the image from the mirror, and readings can be taken on the graduated circle for each position of coincidence. The angle of rotation between the settings of any two images will give the crystal angle between the two reflecting faces. Readings are taken on the graduated circle for the settings of all the faces in the zone and the interfacial angles are deduced; the crystal is then reset for another zone, and in this way values can be obtained for the angle between any two faces on a crystal. The degree of accuracy obtainable varies with the refinement of the instrument and also with the type of crystal face; some faces give a bright, well-defined image, some a double image, and others a blur of light which makes accuracy of setting impossible. In the latter case especially the actual value of the crystal angle must be estimated from the mean value of several determinations*.

Symmetry. From measurement, if not from a more casual inspection, it becomes evident that certain angular values occur more than once amongst the interfacial angles of a crystal, and that crystals possess a certain degree of symmetry, some more, some less. The symmetry may be of different types. In many crystals it is found that every face is matched by a similar parallel face in the antipodal position. Symmetry of this type is known as *centrosymmetry*, each pair of faces being symmetrically disposed about the geometrical centre of the crystal.

Secondly, there is symmetry about a line or *axis of symmetry*. In this case a crystal, when rotated about an axis passing through its centre, presents exactly the same appearance more than once during a complete rotation. If any aspect is presented twice during a rotation through

* For descriptions of more accurate types of goniometer the reader is referred especially to Tutton's *Crystallography and Practical Crystal Measurement*.

360° the axis is said to be one of two-fold symmetry or a *diad axis*; three identical aspects indicate a *triad axis*, and in the same way a crystal may possess *tetrad* or *hexad* axes of symmetry. The lead of a hexagonally shaped pencil represents the hexad axis of the pencil; a vertical line through the summit of the Great Pyramid is a tetrad axis of the pyramid; a match-box possesses three diad axes of symmetry, one parallel to each of its three sets of parallel edges.

The third type of symmetry is symmetry about a plane. *A plane of symmetry* divides a crystal so that each face on the one side is matched by a similar face in a reflected position on the other side of the plane. If a crystal has developed a geometrically perfect form, it is divided by a plane of symmetry so that one half is the mirror image of the other half, and if the plane was replaced by a mirror the one half and its reflection would be exactly similar to the whole crystal. A page half-way through a book represents a plane of symmetry of the book; there is a second plane of symmetry perpendicular to this which would cut the book into two halves perpendicularly to the covers and the back. A match-box has three perpendicular planes of symmetry parallel to its faces; a diagonal plane through opposite edges, cutting it into two equal halves, is not a plane of symmetry, however, for one half is not in the position of the mirror image of the other.

Another type of symmetry which has received increasing recognition by recent workers on crystal structure is the *alternating axis* or *rotation-reflection axis of symmetry*. Similar faces may be found to occur in a crystal in such positions that one can be made to replace another by a simultaneous rotation about an axis and reflection across a plane. A rhomb of calcite affords the best example of this type of symmetry. In fig. 4 a reflection across the horizontal median plane would bring the three lower faces of the rhomb on to the edges between the three upper faces, and a simultaneous rotation of 60° would bring them into coincidence with the upper faces. Thus the vertical axis of the rhomb can be regarded as an axis of composite symmetry, or a six-fold or hexad alternating axis.

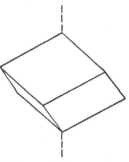

Fig. 4. Calcite rhomb.

Description and Nomenclature. As long as we attempt to refer to any crystal face by describing its relative position in words, we shall run the risk of ambiguity or long-windedness or both. Moreover, by virtue of the Law of Constant Angles, definite geometrical relationships must hold between the various faces, and hence to facilitate mathematical computations as well as the description of spatial arrangement it is advisable to adopt some system of nomenclature for denoting the positions of crystal faces.

The mathematician denotes the position of a point by specifying this position relative to certain arbitrarily fixed lines; two such coordinates fix the position of a point in a given plane, three the position of a point in space. In the same way the crystallographer chooses three fixed lines meeting in a point within the crystal and describes the position of any face relative to these lines, which are known as the *crystallographic axes*. The simplest set of axes to adopt would naturally be three lines at right angles to each other, and these are chosen whenever possible. For many crystals, however, a set of this kind would bear little or no relationship to the general structure of the crystal, and it is found advisable in practice always to choose crystallographic axes which are parallel to well-defined directional lines in the crystal, such as edges, axes of symmetry or the normals to planes of symmetry, even though in some cases one or more of the axes so chosen may be oblique to the others.

The first step, therefore, in the description of a crystal is the arbitrary adoption of a trio of concurrent lines as the crystallographic axes. It is not impossible that different investigators may choose different sets of axes, but for all well-known crystals custom has put its seal of approval on the set of axes which has been found to give the simplest mathematical relationships between the various faces. Arbitrary as the choice of axes is, their position if wisely chosen bears a close relationship to certain fundamental lines in the internal structure of the crystal; this will become evident in the later consideration of the nature of crystal architecture, and leads in practice to a certain degree of inevitability in the selection of axes.

In the general case the plane of any crystal face will cut all three axes, and the ratios of the lengths a, b and c cut off the crystallographic axes by any one face will be a constant for all faces belonging to the same form and in all specimens of the same mineral. After the axes have been selected, the next step in the description of a crystal is to select a face which cuts all three, and by the measurement of angles to determine the relative values of a, b and c for that face. Here again the choice is arbitrary, and only trial can show which is the best face to select in order to facilitate later calculations. The face chosen is known as the *parametral* or *unit plane*, and the three lengths a, b and c are called the *parameters* of the crystal; it must be remembered that these lengths are merely relative, and are constant only as regards their ratios to one another and not in absolute value. In expressing the ratios, the value of b is always written as unity; thus, for example, the parametral constants for the felspar orthoclase are written as

$$a : b : c = 0 \cdot 6585 : 1 : 0 \cdot 5554,$$

and for muscovite, or white mica, as

$$a : b : c = 0 \cdot 5773 : 1 : 3 \cdot 3128.$$

The ratios of the parameters are known as the *parametral* or *axial ratios* for the crystal.

If now the intercepts on the axes are calculated for any other plane—the calculations are made from the measured values of the interfacial angles—it is found that they always bear a very simple relationship to the parameters, being in every case reducible to the parameters multiplied or divided by small whole numbers, 2, 3 or 4, or occasionally by higher numbers.

Fig. 5 shows two faces of a crystal, PQR and HKL, both cutting the three crystallographic axes OA, OB and OC.

If PQR is adopted as the parametral plane, $OP = a$, $OQ = b$ and $OR = c$.

The face HKL can be moved parallel to itself without in any way changing its identity or altering the ratios of its intercepts on the axes: it is found in practice that its position can always be adjusted in this way so that its intercepts on the axis are expressible as ma, nb and pc, where m, n and p are small whole numbers. A crystal face is usually denoted by three indices, and it is from these multipliers m, n and p that the indices of the face are derived. The experimental fact just stated is known therefore as the *Law of Rational Indices*; it is of the utmost importance in the elucidation of crystal structure, and, as will be seen later, is capable of theoretical explanation on the grounds of regular atomic arrangement in crystal building.

The indices of a face are therefore measures of its intercepts on the crystallographic axes, and since the intercepts can always be expressed as integral multiples of the parameters, we may use the integral multipliers as the characteristic indices of the face, although it must always be

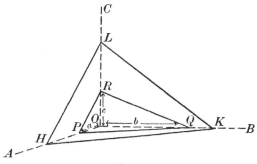

Fig. 5.

remembered that they only express the intercepts in terms of the parameters and not in absolute units of length. In fig. 6 the face *HKL* has been moved parallel to itself into the position *H'K'L'*, where *K'* coincides with *Q*. Its intercepts on the axes now appear as 2*a*, *b* and 2*c*; hence it could be given the indices 2, 1, 2. Following this convention, a face parallel to the axis *OA* would have infinity as its index with regard to that axis; this leads to inconvenience in mathematical calculations, and it has been found more convenient therefore to use as indices numbers proportional to the reciprocals of m, n and p. This is the basis of the *Millerian notation*, so called after W. H. Miller, Professor of Mineralogy in the University of Cambridge, who first brought the system into prominence by adopting it in his classical *Treatise on Crystallography*, published in 1839.

The Millerian indices of the face HKL (fig. 5) would accordingly be h, k, l, where

$$OH : OK : OL = \frac{a}{h} : \frac{b}{k} : \frac{c}{l},$$

and the face is denoted by the symbol (hkl). With the same

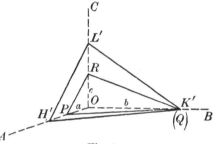

Fig. 6.

notation the symbol for the parametral face would be (111), since its intercepts on the axes are $\frac{a}{1}$, $\frac{b}{1}$ and $\frac{c}{1}$, and hence comes its alternative title of *unit face*. In all cases and for all crystals the law of rational indices states that these Millerian indices are small whole numbers, in practice seldom above four.

In fig. 5 the plane HKL can be moved parallel to itself towards the point O until point H reaches point P; L will then coincide with R and K will be midway between O and Q. This will give the new values of the intercepts of plane

HKL on the axes as $a, \dfrac{b}{2}$ and c, so that the Millerian indices for the plane HKL are (1, 2, 1).

Let us take a real crystal for illustration. Fig. 7 represents a crystal of zircon, a brownish or colourless mineral of the composition $ZrSiO_4$.

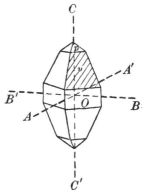

Crystallographic axes have been chosen as shown, the two horizontal axes AOA' and BOB' being perpendicular to two vertical planes of symmetry which bisect the crystal through opposite edges, and the third axis COC' being parallel to the vertical edges in the crystal. There is obviously a close relationship between the structure of the crystal and the axes selected.

Fig. 7. Crystal of zircon.

The two faces marked p and u intersect all three axes, the face u being more abruptly inclined to OC. In fig. 8

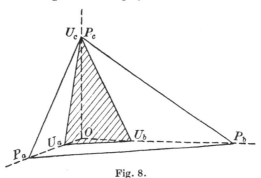

Fig. 8.

they are shown apart from the rest of the crystal, and their relative intercepts on the axes are more evident. If these intercepts are OP_a, OP_b, OP_c for the face p, and OU_a, OU_b and OU_c for the face u, and if one face is moved so that $OP_c = OU_c$, then measurement of the crystal angles shows that

$$OU_a = \frac{OP_a}{3} \quad \text{and} \quad OU_b = \frac{OP_b}{3}.$$

Suppose, then, that we adopt plane p as the unit plane (111); we have for the ratio of the parameters:

$$a : b : c = OP_a : OP_b : OP_c,$$

and the indices of plane u will consequently be (331).

On the other hand, if we choose u as the unit plane (111), the parameters are given by OU_a, OU_b, OU_c, and the indices of plane p become (113). There is no *a priori* reason to choose one face rather than the other as unit plane, but p is the face usually adopted since this leads to a simpler set of symbols for the other faces of the crystal.

When once the directions of the three crystallographic axes are fixed and the values of the parameters a, b and c deduced, the intercepts on the axes of any face (hkl) can be calculated, being $\frac{a}{h}$, $\frac{b}{k}$ and $\frac{c}{l}$, and we can find the position of all possible faces that can occur in the crystal by giving various small integral values to h, k and l. Or, conversely, knowing the interfacial angles we can deduce the indices of any face by the use of comparatively simple mathematical formulae.

In describing a crystal certain constants are always quoted: the angles between the crystallographic axes OA, OB and OC are denoted by α, β, and γ, and in cases where any of these angles are not 90° the actual values are given. These angles and the axial ratios $\frac{a}{b}$, $\frac{c}{b}$ serve to define absolutely the position of the parametral plane, and hence also the position of any other plane whose indices are known; the values of α, β, γ, $\frac{a}{b}$ and $\frac{c}{b}$ are known as the *elements of the crystal*.

The degree of simplicity of these elements will depend upon the symmetry of the crystal. A crystal possessing no symmetry at all will have its crystallographic axes oblique to each other and its parametral ratios fractional. As the degree of symmetry increases, the relations between the axes become simpler until when a crystal shows a high degree of symmetry the axes become mutually perpendicular and finally all equal and therefore interchangeable. In this case the axial ratios are both unity, and a, β and γ are all 90°. This is the symmetry found in crystals of the cubic system (see p. 33).

It has become customary to talk about the relative lengths of the crystallographic axes; by this is really meant the lengths of the intercepts on the axes of the parametral plane. Thus we say that in the cubic system the three axes are perpendicular and equal, while in the system showing least symmetry, the triclinic or anorthic system, the axes are said to be all inclined and all unequal.

There are certain conventions observed as to the orientation of the crystallographic axes. The crystal is always represented with the axis COC' vertical, the positive direction OC being upward; BOB' is taken as running from side to side, the positive direction OB being to the right, and AOA' runs forwards and backwards, the forward direction OA being positive. If the axes are rectangular, and if OA and OB are unequal, the longer is assumed to be OB and is known as the *macro-axis*, while OA is known as the *brachy-axis*. If one axis is inclined, the other two still being at right angles, that axis is adopted as OA, and it is always taken as sloping forward and downward: in this case OA is known as the *clino-axis* and OB as the *ortho-axis*.

Three indices enclosed in plain curved brackets denote a crystal face, but three indices enclosed in curly brackets, such as {110}, denote collectively all the faces belonging to the form of which the face (110) is a member*.

The indices of a face are sufficient to denote its approximate position on the crystal even if the parameters are not known. For example, a face having 0 for two of its indices must be parallel to the two corresponding axes, and its position in the crystal is therefore absolutely fixed. A face parallel to the vertical axis only is known as a *prism face*; this has indices of the type (hk0). A face parallel to the vertical axis and to one of the other axes is known as a *pinacoid* (Gr. πίναξ, a board); there are usually only two pinacoids in a form, although in the tetragonal system the vertical tetrad axis makes the two forms {100} and {010} exactly similar. The two faces of the form {001}, parallel to the two axes which are not vertical, are known as basal pinacoids. Fig. 9 shows a crystal made up of two macro-pinacoids (100) and ($\bar{1}$00), so-called because they are parallel to the macro-axis BOB', two brachy-pinacoids (010) and (0$\bar{1}$0), and two basal pinacoids (001) and (00$\bar{1}$). Similarly the ortho-axis and the clino-axis when present lend their names to the pinacoids that are parallel to them.

---

* If a face cuts an axis on the negative side of o, its index with regard to that axis is a negative quantity and the negative sign is written above it; thus two parallel faces on a crystal are denoted by (hkl) and ($\bar{h}\bar{k}\bar{l}$).

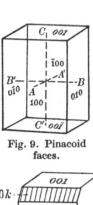

Fig. 9. Pinacoid
faces.

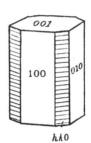

$hk0$

Fig. 10. Pinacoid
and prism faces.

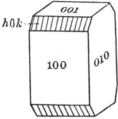

Fig. 11. Pinacoids and
domes.

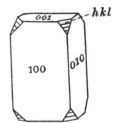

Fig. 12. Pinacoids and
pyramids.

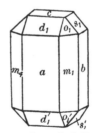

Fig. 13. Crystal of olivine.

$a = (100)$—pinacoid	$d_1 = (h0l)$—dome
$b = (010)$— ,,	$d_1' = (h0\bar{l})$— ,,
$c = (001)$—basal pinacoid	$s_1 = (0kl)$— ,,
$m_1 = (hk0)$—prism	$s_1' = (0k\bar{l})$— ,,
$m_4 = (h\bar{k}0)$— ,,	$o_1 = (hkl)$—pyramid
	$o_1' = (hk\bar{l})$— ,,

The truncation of a vertical edge between two adjacent pinacoidal faces gives rise to a prism face, which is parallel to the vertical axis only (fig. 10). The truncation of a horizontal edge between a pinacoid and a basal pinacoid gives rise to a face of the type $(h0l)$ or $(0kl)$, which is parallel to one of the axes only, and which is known as a *dome* from its resemblance to a roof (fig. 11). Finally, a face cutting all three axes and represented by the general indices (hkl) is known as a *pyramid face* (fig. 12). In fig. 13 these four types of faces all occur together in a typical crystal of the mineral olivine.

REPRESENTATION OF CRYSTALS

Drawing of Crystals. By convention a crystal is always drawn as if it was slightly to the left of, and a little below, the eye of the observer, so that in addition to the front faces those to the right-hand side and on top are also seen. Moreover, lines parallel in the actual crystal appear parallel in the drawing, so that all zonal relationships are clearly preserved. This method of *clinographic projection* has the additional advantage of producing a perfectly natural-looking drawing of the crystal; the *orthographic projection*, on the other hand, which is comparable with the plan and elevation of architectural drawings combined in one diagram, is entirely diagrammatic in appearance, although it gives a clear idea of the crystal habit.

In an accurate clinographic drawing of a crystal the A and B axes are represented as if rotated horizontally $18°\ 26'$ ($\tan^{-1}\frac{1}{3}$) to the left, and the C axis, although it appears vertical, is treated as if it was tilted forward through $9°\ 28'$ ($\tan^{-1}\frac{1}{6}$). The axes are constructed by a special method in accordance with this convention; the parametral lengths are laid off along them, duly modified by perspective, and the crystal faces are then built up on this axial skeleton. For details of the method the reader is referred to Tutton's *Crystallography and Practical Crystal Measurement*, or Rogers's *Study of Minerals and Rocks*.

For ordinary freehand work it is enough to bear in mind the conventional positions of the axes and their relative lengths in any specific case, and to keep co-zonal edges in the crystal accurately parallel in the drawing.

Stereographic Projection. A drawing of a crystal, however carefully made, is not entirely satisfactory. It can at best only show one-half of the faces present, and their shape and size are modified by perspective. Still more unfortunately, a drawing entirely fails to show with any accuracy the size of interfacial angles; it accentuates the habit of the crystal rather than its form. Accordingly methods have been invented to give a representation of a crystal in which non-essentials are eliminated and the important features of form are depicted with mathematical accuracy. Only one method will be described here—that of *stereographic projection*.

As the important characteristic of a crystal face is its position with regard to other faces, it can be represented by a fixed plane of unlimited extent, or more simply by the normal drawn to that plane. Hence all the faces of a crystal can be represented by a series of lines radiating from a point at the centre of the crystal, each being perpendicular to the plane of a crystal face. Suppose these normals to cut the surface of a sphere from whose centre they radiate; the point of intersection of a normal with the sphere may be taken to represent a crystal face, and the whole pattern of points thus formed will be an accurate projection of the crystal on the spherical surface. Two crystal faces set at a large angle to each other will have the same large angle between their normals, so that they will be represented by two points wide apart on the sphere. Two faces at right angles will be represented on the sphere by two points bearing the same relation to each other as the north pole of the sphere and a point on the equator. Hence the actual linear distance measured along a great circle of the sphere between two projected points is an accurate measure of the interfacial angle, which is, in fact, the angle subtended at the centre of the sphere by the normals through the points.

But a pattern on a spherical surface is not of much use for graphical representation, and a second process of projection is added; the points on the sphere are projected on to an equatorial plane, which is chosen as having a definite orientation with regard to the crystal, being generally the plane of two of its crystallographic axes. For this process the following construction is used.

Suppose that the eye is placed at the south pole of the sphere, and suppose that a transparent plane passes through the sphere equatorially. All points on the northern hemisphere will be seen through the transparent plane, and if a line is drawn from the eye to each of these points its intersection with the plane will give its projected position on the plane. The projection of a face on the plane is known as the *pole* of that face.

Fig. 14 shows both stages of the projection. A crystal is seen in outline surrounded by a sphere, each with O at its centre. Normals OB, OS_1, OC, OS_2, ... have been drawn from the centre perpendicular to the faces of the crystal, cutting the sphere in B, S_1, C, S_2, ... Next, lines are drawn from these points of intersection to C', the south pole of the sphere, and these lines cut the equatorial plane in B, s_1, $O, s_2, \ldots$; then $B, s_1, O, s_2, \ldots$ are the poles of the crystal faces on the equatorial plane. It is obvious that the pole B of the vertical face on the right of the crystal lies on the circle which bounds the projections of all points in the upper hemisphere; this is equally true of the poles of all faces perpendicular to the plane of projection. The pole of a face in the lower hemisphere will lie outside this circle, as is the case with the pole s_1', and, as S_1' approaches C', s_1' travels farther and farther from B. This would make the projection a clumsy business in practice; planes in the lower hemisphere are therefore treated differently and are projected on to the equatorial plane in the posi-

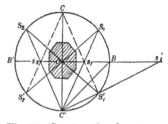

Fig. 14. Construction for stereographic projection.

tions they would assume if viewed from the north pole C. In this way the projections of all possible faces of the crystal will lie within the circle of projection. Poles of faces in the upper hemisphere are represented by dots, those of faces in the lower hemisphere by little rings. If, as often happens, the plane of projection chosen is a symmetry plane of the crystal, every dot will coincide with a ring, since every face above the plane has another face as its mirror image below the plane. In fig. 14 the pole of S_1' will coincide with the pole of S_1, and the pole of S_2' with the pole of S_2.

Let us, for illustration, take the projection of the crystal shown in fig. 13 and also represented in profile in fig. 14. By measurement we find that o_1 is co-zonal with a and s_1 and also with b and d_1 and with m_1 and c.

If the crystal is set up so that the normal to face C is
vertical, the faces in zone a, m_1, b, ... will also be vertical,
and their poles will all lie on the circumference of the
bounding circle, or *primitive circle*, in the projection
(fig. 15). The pole of c will lie at the centre of the primitive
circle, and those of the faces d, s and o will lie within the
circle in the positions shown. The basal plane c' at the

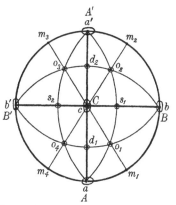

Fig. 15. Stereogram of olivine crystal.

bottom of the crystal will be represented by a ring coincident with the dot c, and similarly each dot on the primitive
circle will be coincident with a ring representing a similar
face on the under side of the crystal.

It is quite obvious that the angle between faces a and
m_1 is given by the angular distance between the points a
and m_1 in the projection. It is less evident, but none the
less true, that the angle between any other two faces of
the crystal can be measured by their distance apart in the
projection.

The angular distance of s_1 from c is represented by cs_1 on the stereogram. Referring to fig. 14, it is evident that the corresponding distance os_1 is equal to $r \tan OC'S_1$, where r is the radius of the sphere and of the circle of projection. But $\angle OC'S_1 = \frac{1}{2} \angle COS_1$ at the centre; this latter angle is contained by the normals to the faces c and s_1, and so is equal to the crystallographic angle between c and s_1. If then any face makes an angle θ with face c, the distance of its pole from c in the projection is $r \tan \dfrac{\theta}{2}$. Special pro-tractors are constructed which give the angular distance from c of any point within the primitive circle, the extreme distance from c to the circumference of the circle representing 90°, and with the aid of these it is possible to plot on the projection any face whose position on the crystal is known, or, conversely, to determine the angular distance between c and any face represented on the projection.

The zonal relationship between faces shows up very clearly in the stereographic projection. The poles of the vertical, prismatic zone a, m_1, b, ... lie on the primitive circle, while poles of the zones a, d_1, c, d_2, a', ... and b, s_1, c, s_2, b', ... lie along the two perpendicular diameters. Zones m_3, o_3, c, o_1, m_1 and m_2, o_2, c, o_4, m_4 are represented by diameters intermediate between these two, while still another zone a, o_1, s_1, o_2, a', ... appears on the projection as the arc of a circle lying intermediate between the dia-meter zone a, d_1, c, d_2, a', ... and the circumference zone a, m_1, b, m_2, a'. In the same way every other zone of the crystal would be represented either by a diameter or by an arc of a circle which cuts the primitive circle at the ends of a diameter; such an arc is known as a *great circle* of the pro-jection. Hence by drawing diameters or great circles it becomes evident at once which faces are co-zonal, or, conversely, if the interfacial angles of a zone have been measured and the positions of two of its faces on the projection are known, the positions of the others can be plotted without difficulty along the diameter or great circle running through these two points.

The stereographic projection also brings into prominence the symmetry of the crystal. Centro-symmetry is indicated by each dot on the projection having a ring equidistant from the centre, the dot and the ring lying on the same diameter; thus in fig. 15 the dot at o_1 is paired by the ring at o_3, the dot at s_1 by the ring at s_2. A dot on the primitive circle will of course be paired by a dot at the other end of the diameter.

An axis of symmetry is represented by a line passing through the centre of the projection and through a pair of parallel crystal faces—or possible faces. Thus in fig. 15 aa', bb' and the line through c perpendicular to the plane of projection are all diad axes of symmetry, for every point swung through 180° about any one of these lines would coincide with a point on the other side of the line. In making a stereogram it is customary to mark the points of emergence of a symmetry axis by distinguishing symbols; $\bigcirc$ denotes the end of a hexad axis, $\square$ denotes a tetrad, $\triangle$ a triad and 0 a diad axis.

A plane of symmetry in the crystal cuts the sphere of projection in a great circle, just as all the normals to a zone of faces cut the sphere in a great circle. Hence in the projection a symmetry plane, like a zone, appears as a diameter or a great circle. In fig. 15 the primitive circle is a plane of symmetry, for every dot on the upper side of the plane has a ring as its mirror image on the under side. Corresponding to every pole on the right of diameter aa', there is a similar pole on the left, and the same is true of poles above and below diameter bb'; aa' and bb', therefore, are planes of symmetry. The diameters m_1m_3, m_2m_4 are not planes of symmetry, however, though they would become so if each bisected the angle between aa' and bb'.

Faces whose poles fall on the primitive circle will all be parallel to the axis running through the centre of the primitive circle and perpendicular to its plane. If, as is very common, the crystal is oriented for projection so that this axis is the vertical axis C, all the poles on the primitive circle will be prisms and their indices will be of the type $(hk0)$. If the A axis lies along the up-and-down diameter of the primitive circle, and the B axis along the side-to-side diameter, faces of the type $(0kl)$ will all lie along the side-to-side diameter, and those of the type $(h0l)$ along the up-and-down diameter.

It is also useful to remember in dealing with faces belonging to a zone, that is to say, with faces whose poles lie along a diameter, the primitive circle or any great circle of the projection, that the sums of like indices of two faces give the indices of a third face intermediate between the first two. $(h_1 + h_2, k_1 + k_2, l_1 + l_2)$ is a face on the same zone as $(h_1k_1l_1)$ and $(h_2k_2l_2)$ and intermediate between them. More generally, if (HKL) is a face intermediate between the faces $(h_1k_1l_1)$ and $(h_2k_2l_2)$, then

$$H = x_1h_1 + y_1h_2,$$
$$K = x_2k_1 + y_2k_2,$$
$$L = x_3l_1 + y_3l_2,$$

where x_1, x_2 and x_3, and y_1, y_2 and y_3 are small integers.

It should be clear without further proof that the stereographic projection presents a very complete and mathematically correct picture of a crystal and, after a little practice has facilitated its use, it becomes of very great assistance in the description and elucidation of crystal morphology. Moreover, with the help of a few simple formulae for the solution of spherical triangles, if certain interfacial angles are known it is possible to deduce others which have not actually been measured on the goniometer. Or, given various interfacial angles, the ratio of the parameters can be found. This procedure is outside the scope of this short book, however; full details will be found in the standard text-books on crystallography. After the description of the crystal systems three actual examples will be worked out to show the construction of a stereogram and the possibility of deducing axial ratios and indices in simple cases without the use of any formulae.

CHAPTER II

THE CLASSIFICATION OF CRYSTALS

The various elements of symmetry that may be present in a crystal have already been described, and usually several of these are combined in any one crystal form. Not all combinations are possible; on the other hand the presence of certain elements involves the presence of others also. For example, the presence of two diad axes at right angles

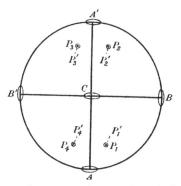

Fig. 16. Stereogram showing three diad axes
and three planes of symmetry.

demands the presence of a third perpendicular diad axis; this is clear from a study of fig. 16 which shows the stereogram of a crystal having two diad axes AA' and BB'. A face P_1 must be repeated at P_2 by reason of the diad axis BB', and at P_4 by reason of the diad axis AA', and again P_2 or P_4 must be repeated at P_3 by the operation of one or other of these axes. This leads to a combination of four like faces, and their symmetry shows that there must be a third diad axis through C perpendicular to the plane of projection.

Again, if to these elements of symmetry is added a centre of symmetry, each face will be paired by a parallel face on the other side of the crystal. P_1 will be paired by a face P_3' whose pole will be directly below P_3, P_2 by P_4', and so forth. Obviously now there are three planes of symmetry, one coincident with the plane of projection, and two perpendicular planes through AA' and BB', and these planes are the necessary corollary to the presence of three perpendicular diad axes and a centre of symmetry.

Given certain minimum elements of symmetry, it is always possible to find what further elements they involve, and it is a task for the geometer to determine the total number of different combinations possible. This was done as early as 1830 by Hessel, a German mineralogist, although at that time no crystals were known which exhibited the symmetry of several of the types enumerated. Bearing in mind the characteristics of crystal morphology, the occurrence of two-fold, three-fold, four-fold or six-fold, but not of five-fold, axes of symmetry, and the law of rational indices, it is possible to enumerate *thirty-two classes of crystal symmetry*, one class exhibiting no symmetry whatsoever, and representatives of every class except one are now known amongst natural minerals or crystals formed in the laboratory.

These classes fall naturally into groups or systems, each group being characterised by a definite set of crystallographic axes. Some crystallographers recognise seven such groups, others prefer to reduce the number to six, regarding one group, the rhombohedral system, as a subdivision of the hexagonal system. This latter system is unique in having four instead of three crystallographic axes assigned to it, three in one plane and a fourth perpendicular to them which is also a hexad axis of symmetry. Mathematically, of course, three co-planar axes are more than is necessary, but, since there is six-fold symmetry about the vertical hexad axis, the choice of three rather than two horizontal axes provides a more natural set and one which leads to similar indices for faces of the same form. The rhombohedral system can also be referred to four axes of this type, or to a set of three equal axes making equal angles with one another which are not right angles. In every other system a set of three axes is chosen bearing definite relationships to the edges of the crystals concerned.

Each system includes a certain number of crystal classes, each class having its own characteristic symmetry. The classes are named in various ways; according to one system a class derives its name from one of the best known crystals which display that symmetry; a more systematic method is to call the class by the name of the geometrical solid figure which results from repeating one face of the form {hkl}— known as the *general form*—in accordance with the symmetry. Given, for example, a class characterised by one tetrad axis and one symmetry plane perpendicular to that axis, the face (hkl) would be multiplied by the axis to give four pyramidal faces symmetrical about the axes, while the plane would demand four more mirror-image faces on its under side. The resulting solid would be a tetragonal bipyramid, and hence this class would be known as the bipyramidal class.

On the end-paper of the book is given a table showing the seven crystal systems, their characteristic crystallographic constants, the elements of symmetry essential to each system and the number of classes each contains. On the page opposite to the table is reproduced the simplest crystal belonging to each system, composed of the forms {100}, {010} and {001}—that is to say, of faces each parallel to two crystallographic axes, so that the axes themselves are parallel to the three sets of edges of the crystal. With hexagonal axes the fundamental crystal is composed of the forms {10Ī0} and {0001}, and with rhombohedral axes it consists of the six faces of the form {100}.

The separate systems will now be described in further detail.

CUBIC OR REGULAR SYSTEM

Characteristic symmetry. Three rectangular diad or tetrad axes, and four triad axes (fig. 17).

Crystallographic axes. Three rectangular, equal axes, coincident with the diad or tetrad axes of symmetry.

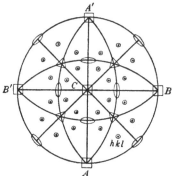

Fig. 17. Stereogram of normal class.
Hexakis-octahedron.

Classes. There are five classes included in the system, in some of which the symmetry is so much reduced that it is difficult to realise at first sight that the crystal has much in common with the *cube*, which gives its name to the system (end-paper, fig. *a*).

For purposes of stereographic projection, the crystal is set up with the *C* axis vertical, the *B* axis running horizontally to the right and left, and the *A* axis running horizontally from front to back. In the stereogram *AA'* and *BB'* lie in the plane of projection, while *CC'* runs perpendicularly through the centre of the stereogram.

CLASS I. HEXAKIS-OCTAHEDRAL OR NORMAL CLASS

Symmetry. The maximum symmetry possible:

> three tetrad axes coincident with the crystallographic axes;
>
> four triad axes equally inclined to the tetrad axes;
>
> six diad axes bisecting the angles between the tetrad axes;
>
> three planes perpendicular to the tetrad axes;
>
> six planes bisecting the angles between these three primary planes.

Fig. 17 shows the elements of symmetry represented on a stereogram, together with the forty-eight faces of the general form {*hkl*}. In this and in the stereograms of other classes planes of symmetry are marked with a heavy line.

Examples. Fluor spar, CaF_2;

> galena, PbS;
>
> analcime, $NaAl(SiO_3)_2H_2O$;
>
> garnet, $(Ca, Mg, Mn, Fe)_3(Al, Fe, Cr)_2(SiO_4)_3$.

This class includes the most regular polyhedral forms, the cube {100}, the octahedron {111} and the rhombic dodecahedron {110}. These forms are all shown on a stereogram in fig. 18.

(*a*) The *cube* is formed by the six faces of the form {100}; it is worth while to detect and mark on a small model of a cube the nine planes and thirteen axes of symmetry characteristic of the class. Simple cubes are often found in crystals of rock salt, fluor spar and galena.

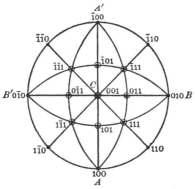

Fig. 18. Stereogram showing cube, octahedron and dodecahedron.

(*b*) The *octahedron* is formed by the eight faces of the form {111}. On the stereogram these coincide with the ends of the four triad axes; in the solid figure these axes emerge perpendicular to the faces of the octahedron, while the tetrad axes and the crystallographic axes emerge from the apices. The octahedron can be regarded as being derived from the cube by cutting off the eight cube corners in a symmetrical fashion (fig. 19); each face is an equilateral triangle and all are equally inclined to each other. Examples are found amongst crystals of spinel ($MgAl_2O_4$) and magnetite (Fe_3O_4).

(c) Another less familiar but important form is the *rhombic dodecahedron*, a polyhedron consisting of twelve similar faces, each rhombic in shape. It is made up of the twelve faces of the form {110}; as the indices imply, each face is parallel to one crystallographic axis, and

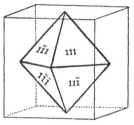

Fig. 19. Octahedron.

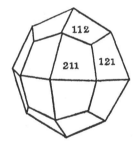

Fig. 20. Rhombic dodecahedron.

Fig. 21. Hexakis-octahedron.

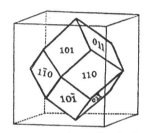

Fig. 22. Icositetrahedron.

equally inclined to the other two. It is the figure produced by truncating symmetrically the twelve edges of a cube until the cubic faces are wholly destroyed (fig. 20). In this case again it affords good practice to find the positions of the nine planes and thirteen axes of symmetry.

The mineral garnet commonly crystallises in good

dodecahedra, and dodecahedral faces also occur on very many other crystals.

(d) The nature of the *general form* which gives its name

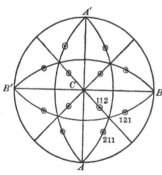

Fig. 23. Triakis-octahedron.

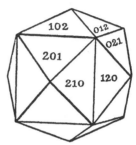

Fig. 24. Tetrakis-hexahedron.

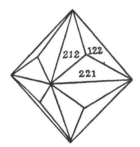

Fig. 25. Stereogram of icositetrahedron.

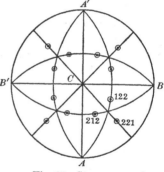

Fig. 26. Stereogram of tetrakis-octahedron.

to the class now becomes apparent. Suppose that in the stereogram the face (111) moves to one side so that its indices are no longer equal but take the general form (*hkl*); its position is shown in fig. 17. The triad axis through (111) will reproduce (*hkl*) in two other positions, and the plane of symmetry through (111) will multiply these to six positions symmetrically disposed about (111). By the action of other planes of symmetry this group of six will be repeated round the other seven faces of the octahedron; hence we shall have a figure with eight groups of six faces, i.e. a *hexakis-octahedron*. Fig. 21 shows the form of this when h, k and l have the values 3, 2 and 1 respectively. Other hexakis-octahedra can be formed by giving different values to h, k and l, but that of the form {321} is the most common. Each face is similar to every other, each having three unlike edges, and the angle over any one type of edge is constant throughout the crystal.

(*e*) When two of the indices are the same, the number of faces of the solid figure is reduced from forty-eight to twenty-four. If the two equal indices are less than the third, the figure produced is known as an *icositetrahedron*; if they are greater than the third, the resulting figure is a *triakis-octahedron*. Fig. 22 shows the icositetrahedron {211} and fig. 25 shows its stereogram. Each face corresponds to the fusion of two adjacent faces of the hexakis-octahedron; for example, the faces (321) and (312) of the latter become the face (211) of the icositetrahedron, and each face has two dissimilar pairs of edges. Garnet affords examples of icositetrahedral crystals, and also the colourless mineral leucite.

Fig. 23 shows the triakis-octahedron {221}, the most common form, and its stereogram is shown in fig. 26. It can also be regarded as the hexakis-octahedron in which adjacent pairs of faces have become united, (321) and (231) becoming (221), or again it can be regarded as an octahedron in which each face has been replaced by three similar faces, each an isosceles triangle.

(*f*) Finally, if one index of the general form (*hkl*) becomes zero, the number of faces is again reduced to twenty-four, and the figure form is a *tetrakis-hexahedron*, or, in other words, a hexahedron or cube in which each face is replaced by a low pyramid of four similar and equally inclined faces, each of these being an isosceles triangle. The simplest, made up of the form {210} (fig. 24), is exhibited sometimes by crystals of fluor spar; other forms such as {310} or {320} also exist.

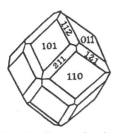

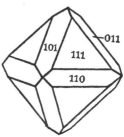

Fig. 27. Garnet showing dodecahedral and icositetrahedral faces.

Fig. 28. Spinel showing octahedral and dodecahedral faces.

Cubic crystals belonging to this normal class may exhibit various combinations of the forms described; garnet, for instance, may show large dodecahedral faces of the form {110} separated by narrow icositetrahedral faces of the form {211} (fig. 27), and the octahedra of spinel are often modified by faces of the form {110} replacing the octahedral edges (fig. 28).

CLASS II. HEXAKIS-TETRAHEDRAL OR TETRAHEDRITE CLASS

Symmetry. Four triad axes as in Class I;
three diad axes coincident with the crystallographic axes;
six diagonal planes of symmetry.

Examples. Tetrahedrite, Cu_3SbS_3;
sphalerite or zinc blende, ZnS.

In this class there is no centro-symmetry, that is to say, no face has a similar parallel fellow on the opposite side of the crystal. The general form {*hkl*} possesses twenty-four faces (fig. 29); fig. 30 shows the hexakis-tetrahedron {321}.

The cube when it occurs shows diagonal striations which are in perpendicular directions on parallel faces, thus destroying their similitude.

The form {111} will possess only four faces, and will

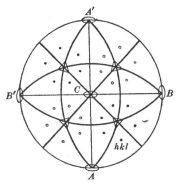

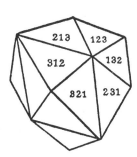

Fig. 29. Stereogram of tetrahedrite
class. Hexakis-tetrahedron.

Fig. 30. Hexakis-
tetrahedron.

produce the solid figure known as a *tetrahedron*. Two tetrahedra are possible, distinguishable theoretically but not in practice. Fig. 31 shows the so-called positive tetrahedron, while fig. 32 shows the corresponding negative tetrahedron. The two forms may occur together, resembling a badly developed octahedron, but the faces of one form are usually smaller than those of the other (fig. 33).

Geometrically a tetrahedron may be regarded as an octahedron whose alternate faces are suppressed, but this conception is not sound from the standpoint of the internal structure of a tetrahedral crystal.

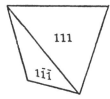

Fig. 31. Positive tetra-
hedron.

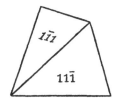

Fig. 32. Negative tetra-
hedron.

Fig. 33. Combination of positive
and negative tetrahedra.

CLASS III. DYAKIS-DODECAHEDRAL
OR PYRITES CLASS

Symmetry. Four triad axes as in Class I;

three diad axes coincident with the crystallographic axes;

three rectangular planes of symmetry, each
perpendicular to a diad axis.

Example. Pyrites, FeS_2.

The cube and the octahedron can both occur in this class, and also the rhombic dodecahedron. The symmetry of the cube is lessened, however, by the presence of striations on the faces (fig. 34), which reduces the symmetry about the crystallographic axes to two-fold symmetry and eliminates the six diagonal planes of symmetry.

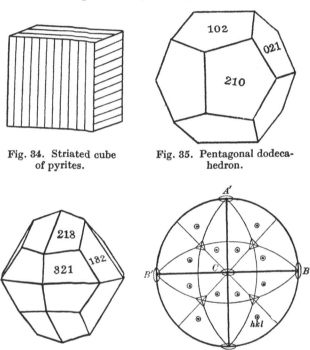

Fig. 34. Striated cube
of pyrites.

Fig. 35. Pentagonal dodecahedron.

Fig. 36. Dyakis-dodecahedron.

Fig. 37. Stereogram of pyrites
class. Dyakis-dodecahedron.

The characteristics of the class are also shown in the form {*hk*0} which in the normal class gave the tetrakis-hexahedron, a figure with twenty-four faces; here it only yields twelve faces and forms the *pentagonal dodecahedron* (fig. 35). Its common occurrence in crystals of pyrites has given rise to its alternative name of *pyritohedron*.

The *dyakis-dodecahedron* (figs. 36, 37) may be regarded as a pentagonal dodecahedron with each face ridged centrally, forming two faces; thus (210) in the dodecahedron may become (321) and (32$\bar{1}$) in the dyakis-dodecahedron.

CLASS IV. PENTAGONAL ICOSITETRAHEDRAL OR CUPRITE CLASS

Symmetry. Three tetrad axes, as in the Normal Class;
four triad axes, as in the Normal Class;
six diad axes, as in the Normal Class.

Examples. Cuprite, Cu_2O;
ammonium chloride, NH_4Cl.

The absence of any plane of symmetry and of a centre of symmetry is indicated by the different orientation of striations or other surface markings on the cubic faces, but cube, octahedron and rhombic dodecahedron each have

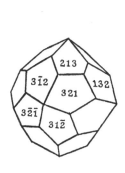

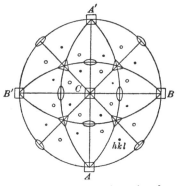

Fig. 38. Left pentagonal icositetrahedron.

Fig. 39. Stereogram of cuprite class. Pentagonal icositetrahedron.

their full number of faces. Not many examples of the class are found, but the few examples known are especially interesting as showing a right-handed and a left-handed form bearing to each other the relationship of an object and its mirror image, a phenomenon known as *enantiomorphism**. Figs. 38 and 39 show the type form of the class and its projection; the indices of the *pentagonal icositetrahedron* may be of any value, and a set as complicated as (896) is quoted by Miers as having been detected in crystals of cuprite.

CLASS V. TETRAHEDRAL PENTAGONAL
DODECAHEDRAL OR ULLMANNITE CLASS

Symmetry. Four triad axes, with dissimilar ends;
     three diad axes coincident with the crystallographic axes.

Examples. Ullmannite, NiSbS;
     barium nitrate, $Ba(NO_3)_2$;
     sodium chlorate, $NaClO_3$.

This is the least symmetrical class of the cubic system. The cube and the dodecahedron {110} have their full number of faces, but the lower degree of symmetry reduces the number of {111} faces to four, so that the octahedron gives place to the tetrahedron. The type form (fig. 40) resembles a tetrahedron in which each face is divided into three irregular pentagonal faces, and the lack of symmetry is obvious both in the solid figure and in its stereogram (fig. 41).

---

* For discussion of this phenomenon of enantiomorphic forms see especially Barker, *The Study of Crystals*, and Tutton, *Crystalline Form and Chemical Constitution.* See also pp. 61 and 62.

To the cubic system belong several important rock-forming minerals, including especially various metals and some of the simpler compounds such as pyrites (FeS_2), galena (PbS), and fluor spar (CaF_2). Cubic crystals can usually be recognised by their similar growth in all three

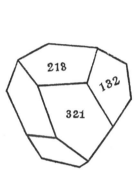

Fig. 40. Left positive tetra-hedral pentagonal dodeca-hedron.

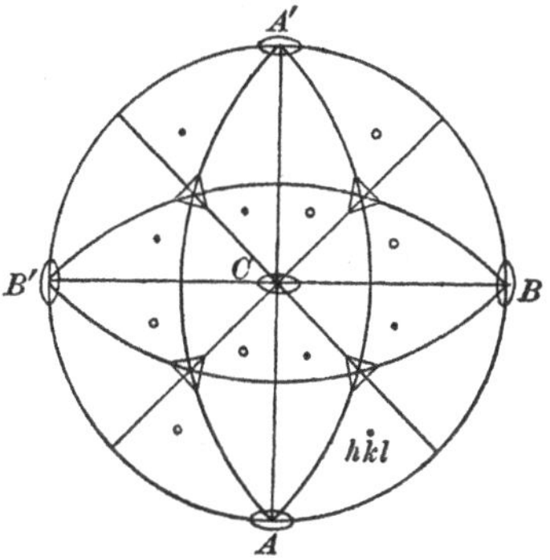

Fig. 41. Stereogram of ullmannite class. Tetrahedral pentagonal dodecahedron.

dimensions, although it must be remembered that regular growth depends upon conditions of environment and may often be prohibited, the crystal becoming elongated or flattened. It is, for example, only by measurement of angles and the consequent detection of triad axes of symmetry that an elongated cubic crystal can be differentiated with certainty from one belonging to the tetragonal system.

TETRAGONAL SYSTEM

Characteristic symmetry. One tetrad axis.

Crystallographic axes. Three rectangular axes, two being equal and the third coinciding with the tetrad axis of symmetry.

Classes. There are seven classes, in which the symmetry varies from the minimum of one tetrad axis, simple or alternating, coinciding with the *C* crystallographic axis, to the fullest possible amount, namely, one tetrad axis, four vertical planes of symmetry at 45° to each other and intersecting in the tetrad axis, a fifth plane perpendicular to these, and consequently also four diad axes at the intersection of this plane with the other four. Fig. 42 shows the

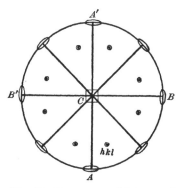

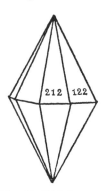

Fig. 42. Stereogram of tetragonal system with full symmetry.

Fig. 43. Di-tetragonal bipyramid.

stereogram of the most symmetrical class with the general form {*hkl*}. The orientation of the crystal for purposes of projection is the same as for the cubic system, the unequal axis being chosen as the vertical axis *C*. The type crystal formed by the four pinacoids {100} with the two basal pinacoids {001} is a square-sectioned rectangular pillar (see end-paper, fig. *b*), and a second rectangular pillar is formed by the prism faces {110}. Since there are no triad axes, {100} is distinct from {001} and the prism {110} from the dome {101}. The pyramid faces {111}, each an isosceles

THE CLASSIFICATION OF CRYSTALS   47

triangle, form a square pyramid, and if the eight faces of
the form are all present they make up what is known as a
bipyramid, which in the tetragonal system represents the
octahedron of the cubic system. Similarly, the equilateral
faces of the cubic tetrahedron are replaced by faces elongated or shortened in the direction of the C axis into isosceles
triangles, the four faces of the form making up what is
known as a *tetragonal bisphenoid* (fig. 45), from its charac-

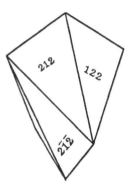

Fig. 44. Tetragonal scalenohedron.

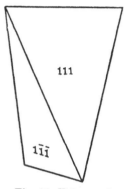

Fig. 45. Tetragonal
bisphenoid.

teristic double-ended wedge-like shape. Just as the tetrahedron can be regarded superficially as derived by the
suppression of the alternate faces of an octahedron, so the
bisphenoid can be regarded as derived from a bipyramid.
Chalcopyrite, or copper pyrites, often shows the sphenoidal
form indicated in fig. 46, the small faces being distinguishable in lustre from the larger faces.

The general form $\{hkl\}$ will consist of sixteen faces forming
a bipyramid with four pairs of faces above and four pairs
below the median plane of symmetry, and this is described

as a *di-tetragonal bipyramid* (fig. 43). If now we suppress alternate pairs of faces, we derive a new figure known as the *tetragonal scalenohedron* (fig. 44), each face being a scalene triangle. Both the bisphenoid and the scalenohedron possess an alternating and not a simple tetrad axis, and obviously do not possess the full symmetry of the di-tetragonal bipyramid. Again, by the suppression of alternate faces in the di-tetragonal bipyramid, we arrive at the two enantiomorphous forms of the *tetragonal trapezohedron* {212}

Fig. 46. Crystal of
chalcopyrite.

Fig. 47. Tetragonal
pyramid.

and {122}, comparable with the trapezohedra of the hexagonal system (fig. 58).

Another interesting class is that in which the tetrad axis has different terminations, as in the tetragonal pyramid of fig. 47. Crystals with this type of axis often exhibit a curious electrical phenomenon, for when they are heated the two ends of this uniterminal axis may become oppositely charged, the effect being reversed on cooling. This phenomenon, which is known as *pyro-electricity*, can, of course, only be exhibited by non-conducting crystals.

This short description should be enough to make the

identification of good tetrahedral crystals an easy matter, and the seven different classes within the system will not be treated in any further detail.

Tetragonal crystals are most commonly prismatic in habit, showing pinacoids and prisms, and often terminated by dome and pyramidal faces, but the pyramidal and tabular habits also occur frequently.

ORTHORHOMBIC OR RHOMBIC SYSTEM

Characteristic symmetry. One diad axis which is the line of intersection of two planes of symmetry, or which is combined with two other perpendicular diad axes.

Crystallographic axes. Three rectangular axes, all unequal, coinciding with the three diad axes when these are present. $a < b \neq c$. The type crystal formed by the pinacoids is still rectangular but shows three different pairs of pinacoids since the parameters are now all unequal (end-paper, fig. c). Convention dictates that a shall be less than b; any of the three axes is chosen as the vertical C axis, and then the larger of the other two is taken as the side-to-side B axis. An upright match-box with the striking surface at the side is an illustration of the simplest rhombic crystal. This inequality of axes has given rise to the prefixes macro- and brachy- (see p. 19), to distinguish between the horizontal axes and between pinacoids or domes parallel to them. For purposes of projection the same orientation of the crystal is used as for the cubic and tetragonal systems.

In the description of a rhombic crystal the three-fold ratio $a : b : c$ is always quoted, b being taken as unity. For example, the axial ratios for the rhombic mineral topaz are quoted as $a : b : c = 0 \cdot 528 : 1 : 0 \cdot 477$, for enstatite ($MgSiO_3$)

as $a : b : c = 0.970 : 1 : 0.571$, and for barytes ($BaSO_4$) as $a : b : c = 0.8152 : 1 : 1.3136$.

There are three classes in the system; in the most symmetrical the crystallographic axes are all diad axes of

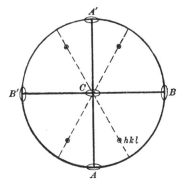

Fig. 48. Stereogram of rhombic system
with full symmetry.

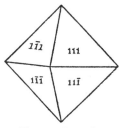

Fig. 49. Rhombic
bipyramid.

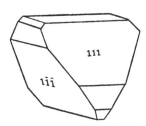

Fig. 50. Rhombic bisphenoid
of tartar-emetic.

symmetry and are also the lines of intersection of three mutually perpendicular planes of symmetry (fig. 48). In the second class there are no planes of symmetry, and in the least symmetrical class only the vertical axis is a diad

axis, and it is also the line of intersection of two perpendicular planes of symmetry. This class is represented by the
mineral hemimorphite, or silicate of zinc, which possesses
a uniterminal vertical axis and shows the phenomenon of
pyro-electricity.

The form {111} is a bipyramid on a rhombic base (fig. 49),
instead of on a square base as in the tetragonal system. The
rhombic bisphenoid is also found, each face being a scalene
triangle; crystals of tartar-emetic (fig. 50) and also of
Epsom salts exhibit this form, which possesses three diad
axes but no planes of symmetry.

Rhombic crystals are common in the mineral world.
While some show the characteristic sphenoidal form, and
others incline to a pyramidal habit, the commonest habit
is perhaps the tabular, although prismatic crystals are also
very often found.

MONOCLINIC OR OBLIQUE SYSTEM

Characteristic symmetry. One diad axis and one plane
of symmetry perpendicular to it, or either one of these
separately.

Crystallographic axes. Three unequal axes, one perpendicular to the other two and coinciding with the diad axis
of symmetry. $a \neq b \neq c$; $\beta \neq 90°$.

By studying the trend of the edges of a monoclinic
crystal it is easy to select two sets of parallel edges at right
angles to one another and a third set at right angles to one
of the first two but inclined to the other. This third set is
adopted as giving the direction of the inclined A axis, and
the crystal is always drawn or described as having the A axis
sloping downwards towards the observer. It will then make
an obtuse angle with the upward vertical axis C, and either

this angle or its supplement is quoted as the value of β. The third axis, being perpendicular to the A and C axes, falls into the usual horizontal side-to-side position of the B axis.

One class of the system possesses a diad axis coincident with the B axis; a second class has instead a plane of symmetry perpendicular to this axis, while the third and most symmetrical class of the system has both.

With the same orientation as for the rectangular systems of axes, the A axis would no longer lie in the plane of the stereographic projection. To avoid unnecessary complications and to show more clearly the size of the important angle β, the plane of the axes A and C is adopted as the plane of projection instead of the plane of the axes A and B.

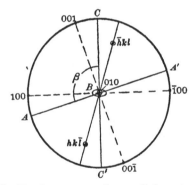

Fig. 51. Stereogram of monoclinic system.

This makes B appear at the centre of the primitive circle; CC' is made to run from top to bottom, and hence AA' takes up a position similar to that shown in fig. 51. The angle ABC then gives the value of β. (100), being parallel to the B and C axes, must have its pole at the western pole

of the primitive circle, taking C as the north pole. The pole
of (010) will lie at the centre of the primitive circle, and
(001), being parallel to the A and B axes, must have its
pole on the primitive circle at an angular distance of 90°
from A. It is clear that the value of β is also given by the
angle between (100) and (001). It is important to remember that the A axis, owing to its tilt, no longer emerges
from the crystal perpendicular to (100), and that this
prevents the face (001) from being perpendicular to the
C axis.

The pinacoids will be of three distinguishable kinds, the
basals {001} inclined to the vertical axis, the ortho-pinacoids
{100} parallel to the ortho-axis B, and the clino-pinacoids
{010} parallel to the clino-axis A. The dome faces (101) and
(011) will be distinguished by the same prefixes, and (101)
will not be of the same form as (011).

The simple crystal defined by the six pinacoidal faces is
like the simple rhombic crystal which has undergone de-
formation by pressure being applied to the edges between
(100) and (001) and ($\bar{1}$00) and (00$\bar{1}$) (end-paper, fig. d).
These pinacoidal faces very often occur in monoclinic
crystals, almost always in combination with dome, prism
and pyramid faces. Often the crystal is elongated in the
direction of one axis, generally parallel to COC', producing
a prismatic habit, but sometimes also in the direction of the
A axis, as in some specimens of the felspar orthoclase, or
parallel to the B axis, as in the mineral epidote. In other
cases the crystals are tabular, as for instance in the case
of gypsum, which shows large clinopinacoids. Very many
important minerals crystallise in monoclinic form, and
with scarcely any exceptions they belong to the class
possessing both an axis and a perpendicular plane of

symmetry. They may occasionally be mistaken for rhombic crystals; for example, orthoclase may form crystals as shown in fig. 52; the faces (001) and (10$\bar{1}$) are almost equally inclined to the vertical, resembling the two brachy-domes of a rhombic crystal. However, there is actually a

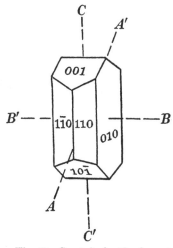

Fig. 52. Crystal of orthoclase.

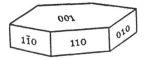

Fig. 53. Crystal of mica.

difference of about 6° in their inclination to the vertical, and they are further distinguished by their difference in lustre and by the presence of one perfect cleavage parallel to (001).

Another monoclinic mineral, mica (fig. 53), has a pseudohexagonal appearance. It has a tabular habit, with large basal pinacoids and a development of prism and vertical pinacoidal faces at very nearly 60° with each other. The angle β varies in different varieties of mica, but it always approximates to 90°, and at first sight a mica crystal appears to have a vertical hexad axis, the actual presence of which can however be disproved by careful measurement.

TRICLINIC OR ANORTHIC SYSTEM

Characteristic symmetry. Centro-symmetry or none at all.

Crystallographic axes. Three axes, unequal and inclined. $a \neq b \neq c$; α, β and γ vary.

The crystal faces may occur in parallel pairs, that is to say, the crystal may have centro-symmetry; otherwise the system possesses no symmetry whatever. Crystallographic axes can be chosen quite arbitrarily from sets of parallel edges exhibited by the crystal, and three are selected which are as nearly perpendicular as possible. By convention the C axis is placed vertically, the A axis running backwards and forwards and sloping down towards the observer, and the B axis running from left to right but usually not horizontally (end-paper, fig. *e*). The plane perpendicular to C is chosen as the plane of projection, so that in the stereogram C will be at the centre of the primitive circle, A will lie on the north and south diameter but within the primitive circle and in the lower hemisphere, while B is within the right half of the primitive circle. The pinacoids (100) and (010) lie on the primitive circle but not at 90° apart; the basal (001) is near but not at the centre (fig. 54).

A triclinic crystal showing only pinacoid faces has a most unhappy, deformed appearance. The diagonal flattening of a rhombic crystal to produce a monoclinic crystal, followed by a second diagonal pressure on the edges (010) to (001) and (0$\bar{1}$0) to (00$\bar{1}$), to produce flattening in a second direction, would result in an asymmetric solid which would have as its sole claim to symmetry the occurrence of pairs of similar and parallel faces. Triclinic crystals do not in reality owe their existence to any such process, but their appearance suggests this mode of origin, and it serves also to describe the relationship between the systems.

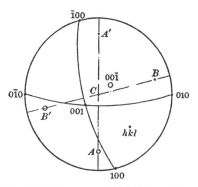

Fig. 54. Stereogram of triclinic system.

Several rock-forming minerals belong to the triclinic system, and usually their crystals are found to possess centro-symmetry. Their departure from monoclinic symmetry is often very small, and triclinic minerals which by chemical composition form a graded series will form a graded series also with regard to their inter-axial angles; this is the case with the plagioclase felspars which provide the most important examples of common triclinic minerals. Their crystals resemble monoclinic forms and exhibit especially pinacoids, prisms and pyramid faces, the crystals showing no pronounced habit of growth.

HEXAGONAL SYSTEM

Characteristic symmetry. One hexad axis.

Crystallographic axes. Three similar co-planar axes at 60°, A_1, A_2 and A_3, and a fourth unique axis C perpendicular to these three, coincident with the hexad axis of symmetry. The reason for the adoption of four rather than

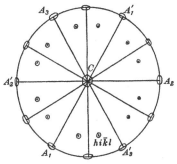

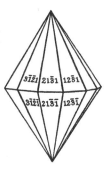

Fig. 55. Stereogram of hexagonal system showing di-hexagonal bipyramid ($hi\bar{k}l$).

Fig. 56. Di-hexagonal bipyramid.

three axes has already been explained (p. 32); the six-fold symmetry about the unique axis calls for a choice of axes exhibiting the same symmetry, and the clumsiness of four-fold indices is overbalanced by the fact that with four axes like faces receive similar sets of indices. Moreover, knowing the intercepts of any face on two of the three similar axes, we can obviously calculate the intercept on the third, and in practice we find that for any face the sum of the indices with regard to the three co-planar axes is equal to zero.

For purposes of description, drawing and stereographic projection, hexagonal crystals are set up with the unique axis vertical, and one of the three similar axes, A_2, is made to run from right to left (fig. 55 and end-paper, fig. f).

Since it is not possible to have a face cutting only one horizontal axis, there are no pinacoidal faces except the two basal planes whose indices are (0001) and (000$\bar{1}$). All faces parallel to the vertical axis are prisms, the most fundamental of which are the faces of the first order prism, fig. 57, comprising (10$\bar{1}$0), parallel to A_2, (01$\bar{1}$0) parallel to A_1, and ($\bar{1}$100) parallel to A_3, together with the other three faces of this form, and the prism faces of the second order which truncate the vertical edges of the first order prisms and are each perpendicular to one horizontal axis. Their indices are of the form {11$\bar{2}$0}.

There are no dome faces; all faces inclined to the vertical axis are pyramids, those of the first order having indices of the form {10$\bar{1}$1} (fig. 57), or more generally {$h0\bar{h}l$}, while those of the second order have indices of the form {$hh\overline{2h}l$}.

In many ways the system is similar to the tetragonal system, the four-fold symmetry of the latter being replaced by six-fold symmetry. Each class within the hexagonal system corresponds to one within the tetragonal system, and to the inevitable vertical hexad axis is added in the various classes six diad axes in the horizontal plane, one horizontal plane of symmetry, giving the typical bipyramidal figure, six planes of symmetry intersecting in the hexad axis, and finally in the most symmetrical class all these additional elements of symmetry are combined together. The general form of this class is the *di-hexagonal bipyramid* (fig. 56). There are no classes characterised by an alternating hexad axis, however; the bisphenoidal and scalenohedral classes are relegated to the rhombohedral system, for a hexad alternating axis is equivalent to a triad axis of a rather special type.

The characteristic forms of the hexagonal system are therefore the hexagonal prism, the hexagonal pyramid and

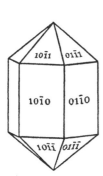

Fig. 57. Hexagonal crystal showing first order prisms and domes.

Fig. 58. Hexagonal trapezohedron.

the hexagonal trapezohedron (fig. 58), derived as in the case of the tetragonal trapezohedron by the elimination of alternate faces of the di-hexagonal bipyramid (fig. 56), the general form $\{hi\bar{k}l\}$ of the most symmetrical class; while the development of large basal faces occurs in crystals of a tabular habit.

RHOMBOHEDRAL OR TRIGONAL SYSTEM

Characteristic symmetry. One triad axis.

Crystallographic axes. Three similar co-planar axes at 60°, and one unique axis perpendicular to them, coinciding with the triad axis. $a = a = a \neq c$.

Rhombohedral crystals are very similar to hexagonal crystals in which alternate faces have been eliminated. Some authorities prefer to include them in the hexagonal system; others prefer to use a special set of rhombohedral axes in dealing with crystals of this type (see end-paper, figs. *g* and *h*). While treating them as belonging to a separate system, we shall still make use of the axes of the hexagonal system, and this will help to emphasise the linkage between the two systems.

The faces tend to appear in groups of three; trigonal prisms and trigonal pyramids are of common occurrence and may be of the first or second order as in the hexagonal system. They will bear similar indices, but there will only be three faces in each form. The trigonal scalenohedron appears, either by itself or in combination with prismatic or other faces.

When, however, the triad axis is also an alternating hexad axis, new forms appear. The form $\{h0\bar{h}l\}$ will consist of six faces, three grouped about each end of the axis, and together they make up the geometrical figure known as a *rhombohedron* (end-paper, fig. *h*), each face being rhombic in shape. If these faces become rectangular, the rhombohedron becomes a cube; except in this special case a rhombohedron is termed acute or obtuse according to the size of the solid angle made by the faces at either end of the triad axis.

The general form {hikl}, combined with an alternating hexad axis, gives a twelve-faced solid figure, a di-trigonal scalenohedron, the faces meeting in a zigzag junction round the middle of the crystal; incidentally this zigzag junction is the same as that made by the median edges of the rhombohedron {h0hl}, so that the two figures bear the relationship shown in fig. 59.

These various forms are of considerable importance practically as well as theoretically, for they are found among the crystals of two of the most important minerals, quartz and calcite. Quartz, possessing three diad axes in addition to the triad axis of the system, forms typically prismatic crystals, closely resembling hexagonal forms and only separable from them superficially by the occasional presence of small pyramidal faces occurring in groups of three (see figs. 63 and 64).

Fig. 59. Rhomb derived from scalenohedron.

The stereogram of the quartz class is shown in fig. 60, in which the position of the general form {hikl} is indicated. The six faces of this form make up a trigonal trapezohedron (fig. 61); a second similar trapezohedron is shown in fig. 62, which is the mirror image of the first. By no manipulation can these two figures be brought into identical positions; they bear to one another the relationship of a right hand to a left hand, and we must therefore expect quartz and other minerals crystallising in this class to appear in right- and left-handed forms (figs. 63 and 64). This is the phenomenon known as enantiomorphism; examples have already been referred to among crystals in other systems showing certain elements of symmetry (see p. 44); it is found to be of special significance when the optical properties of minerals are studied*.

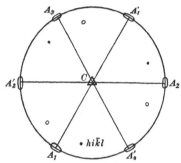

Fig. 60. Stereogram of quartz class
(left-handed crystal).

Fig. 61. Right trigo-   Fig. 62. Left trigonal   Fig. 65. Dog-tooth
nal trapezohedron.   trapezohedron.   spar.

Calcite crystals possess in addition to the triad axis three diad axes of symmetry in the perpendicular plane, and also three planes of symmetry intersecting in the triad axis and

* See Miers, *Mineralogy*, ch. vi, or Tutton, *Crystallography*, ch. xlix.

bisecting the angles between the diad axes. This is equivalent to an alternating hexad axis, and consequently calcite crystals occur commonly in rhombohedra, scalenohedra, the well-known "dog-tooth spar" (fig. 65), and combinations of the two. Another habit is a prismatic form, made

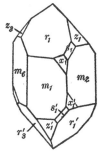

Fig. 63. Right-handed
quartz crystal.

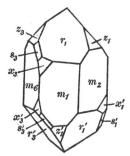

Fig. 64. Left-handed
quartz crystal.

up of long, six-sided prisms capped with sharp pyramidal faces.

Other rock-forming minerals also occur in rhombohedral form, but even without these the presence of quartz and calcite in the rhombohedral system would suffice to make its study of first rate importance to the mineralogist.

PRACTICAL EXAMPLES IN STEREOGRAPHIC PROJECTION

Three examples will be given to illustrate the use of stereograms in connection with crystal measurements and calculations. No use will be made of formulae for the solution of triangles described on a spherical surface, nor of the invaluable anharmonic ratio connecting co-zonal faces; for these, and for any but the most straightforward problems which can be solved by graphical methods and simple trigonometry, the reader is referred to Lewis's *Crystallography* or other standard text-books.

I. *To make a stereogram of the orthorhombic crystal olivine, and to find its axial ratios.*

The crystal illustrated in fig. 13 was measured on the goniometer with the following results:

In the prism zone,

$$am_1 = m_2a' = a'm_3 = m_4a = 24° 57\tfrac{1}{2}',$$

and $$ab = ba' = a'b' = b'a = 90°.$$

In zone $ad_1c, \dots,$

$$ad_1 = d_2a' = a'd_2' = d_1'a = 38° 27',$$

and $$ac = ca' = a'c' = c'a = 90°.$$

In zone $bs_1c, \dots,$

$$bs_1 = s_2b' = b's_2' = s_1'b = 59° 37',$$

and $$bc = cb' = b'c' = c'b = 90°.$$

Faces $m_1, o_1, c, o_3, m_3, \dots$ are found to be co-zonal, and also faces $b, o_1, d_1, o_4, b', \dots.$

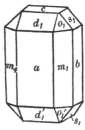

Fig. 13. Crystal of olivine.

These measurements show that the crystal possesses three planes of symmetry parallel to a, b and c, and that these planes intersect in diad axes; there is also centro-symmetry. Hence the crystal shows the fullest symmetry possible in the orthorhombic system.

Let us assume that a is (100), b is (010) and c is (001). In the stereogram a and b will be on the primitive circle at

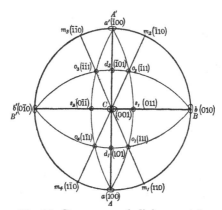

Fig. 66. Stereogram of olivine crystal.

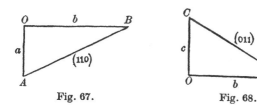

Fig. 67.   Fig. 68.

90° from each other, and c at the centre, with c' (00$\bar{1}$) vertically below and represented by a ring. ($\bar{1}$00) and (0$\bar{1}$0) will be at the other ends of the diameters through (100) and (010).

Faces m_1, m_2, ... will lie on the primitive circle and can be plotted. d_1 and d_2 will lie on the diameter aca'; their positions can be plotted by eye, or, if an accurate stereogram is required, the linear distance of d from c is $r \tan \dfrac{51°33'}{2}$, where r is the radius of the primitive circle (see p. 26).

The positions of s_1 and s_2 along the diameter bb' can be found in a similar way. d_1' and d_2', s_1' and s_2' will be represented by rings round d_1 and d_2, s_1 and s_2 respectively.

Face o_1 is common to the zones co_1m_1 and d_1o_1b; hence its position is given by the intersection of the arc of a circle through b, d and b' with the radius cm_1, and the positions of o_2, o_3 and o_4 can be found in the same way. o_1', o_2', o_3' and o_4' will be represented by rings round o_1, o_2, o_3 and o_4 respectively.

To find the axial ratios, we may choose any plane cutting all three axes as the parametral plane (111), or alternately we can choose two of the three planes (110), (011) and (101), thus fixing separately two of the ratios between the parameters a, b and c.

Let us then take m_1 as (110) and d_1 as (101).

Since (111) is given by the addition of like indices of planes (101) and (010), and also of planes (001) and (110), (111) must be the indices of the plane o_1 which is common to both these zones.

Again, if a great circle is drawn through a, o_1, o_2, and a', it is found to pass through s_1; that is to say, s_1 is the point of intersection of the arcs representing zones (001) and (010), and (111) and (1̄11). The addition of the latter pair of indices gives a plane (011), and this is also given by adding the indices (001) and (010). Hence the indices of s_1 are (011).

In fig. 67, if AB is the trace of plane (110) in the axial plane OAB, $OA = \dfrac{u}{1}$, and $OB = \dfrac{b}{1}$, so that

$$\frac{a}{b} = \frac{OA}{OB} = \tan am_1 = 0\cdot4656.$$

In fig. 68, if BC is the trace of plane (011) in the axial plane OCB, $OC = \dfrac{c}{1}$, and $OB = \dfrac{b}{1}$, so that

$$\frac{c}{b} = \frac{OC}{OB} = \tan s_1 c = 0\cdot5865.$$

Hence $a : b : c = 0\cdot4656 : 1 : 0\cdot5865.$

II. *To make a stereogram of a monoclinic crystal of pyroxene,*
given the axial ratios a : b : c = 1·0921 : 1 : 0·5893, *and*
$\beta = 74° 10'$.

A sketch of the crystal is shown in fig. 69. It exhibits the following forms: $a\,\{100\}$, $b\,\{010\}$, $c\,\{001\}$, $m\,\{110\}$, $p'\,\{10\bar{1}\}$, $u\,\{111\}$ and $v\,\{221\}$.

For purposes of projection we take the B axis perpendicular to the plane of the paper, the C axis vertically up and down in the plane of the paper and the A axis in the same plane making $\angle ABC = 180° - 74° 10'$. $a\,(100)$ and $a'\,(\bar{1}00)$ will lie at the extremities of the horizontal diameter of the stereogram, at 90° from B and from C, $b\,(010)$ and $b'\,(0\bar{1}0)$ will lie at the centre of the primitive circle, and $c\,(001)$ and $c'\,(00\bar{1})$ at the extremities of a diameter perpendicular to the A axis (fig. 70).

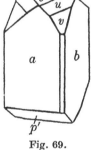

Fig. 69.

In a plane perpendicular to the C axis let OB represent

the B axis, and BA' the trace of plane m (110) (fig. 71).

Then $OA' = OB \tan (100) \wedge (110)$. Now $OB = \dfrac{b}{1}$, and

$OA' = \dfrac{a}{1} \sin 74° 10'$, since the A axis makes with the plane

of the paper an angle of $90° - 74° 10'$. Hence

$$\tan (100) \wedge (110) = \frac{a \sin 74° 10'}{b} = \tan 46° 25'.$$

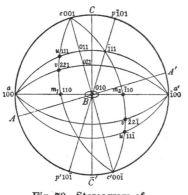

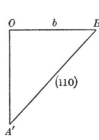

Fig. 70. Stereogram of
pyroxene crystal.

Fig. 71.

Plane (110) is obviously in the zone am_1b, so that m_1 can
be plotted in the stereogram at a distance $46° 25'$ from a,
and m_4 will lie directly below it. Similarly m_2 and m_3 will
be on the same diameter at a distance of $46° 25'$ from a'.

In a similar way the position of (011) can be found from
the ratio $\dfrac{c}{b}$; (011) will lie on the diameter through (010)
and (001).

u (111) is at the intersection of zones (001), (110), and
(100), (011), and hence can be plotted; (1Ī1) will be directly
below it, and (Ī1Ī) and (ĪĪĪ) will be on the diameter through

(111) equidistant from, and on the other side of, **B**.
Similarly, (111) must be a plane common to the zones (011),
($\bar{1}$00), and (001), ($\bar{1}$10); both these zones can be drawn and
the position of ($\bar{1}$11) determined by their point of intersection. Then p ($\bar{1}$01) and p' (10$\bar{1}$) will lie at the ends of the
diameter through ($\bar{1}$11).

Finally, v (221) must lie between (111) and (110) and
between (100) and (021). This latter face must lie between
(010) and (011) in such a position that

$$\tan (001) \wedge (021) = \frac{2c}{b} \sin 74° 10',$$

so that its exact position can be found. Then the zone (100),
(021) can be drawn, and its point of intersection with the
zone (111), (110) will give the position of v (221).

III. *To make a stereogram of a quartz crystal (rhombohedral system), and to find its axial ratios.*

The crystal is shown in fig. 63. Measurement on the
goniometer showed that m_1, x_1, s_1, z_1, r_2 and m_4 belong to
one zone, m_2, s_1 and r_1 to another, m_2, z_1, r_3 and m_5 to a
third, and that $m_1r_1 = 38° 13' = m_2z_1$. The vertical axis is an
alternating hexad axis of symmetry, and there are three horizontal diad axes emerging through opposite pairs of edges
of the prism zone, but there are no planes of symmetry.

To m_1 and r_1 are always given the indices (10$\bar{1}$0) and
(10$\bar{1}$1) respectively. The three equal axes A_1, A_2 and A_3
are plotted in the plane of the paper; the m faces lie on the
primitive circle midway between the ends of the axes, and
the r and z faces lie alternately on diameters through the
m faces and at an angular distance of 38° 13' from the latter.
r_1' will lie vertically below z_1, and z_1' vertically below r_1.

The great circle representing the zone $m_1r_2m_4$ can be drawn; s_1 lies on this circle and also on the circle representing the zone $m_2s_1r_1m_5$, so that the position of s_1 can be

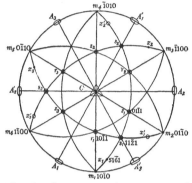

Fig. 72. Stereogram of quartz crystal.

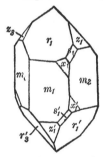

Fig. 63. Right-handed quartz crystal.

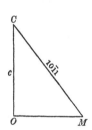

Fig. 73.

located. s_2 and s_3 can be found in the same way. The indices of s_1 are derived by adding p times $(10\bar{1}0)$ to q times $(\bar{1}101)$ (which are the indices of r_2), and also by adding x times $(01\bar{1}0)$ to y times $(10\bar{1}1)$; by taking $p = 2$, and $q = x = y = 1$, we deduce that s_1 has the indices $(11\bar{2}1)$.

x_1 lies on the zone $m_1s_1z_1r_2m_4$ and between m_1 and s_1, but its exact position cannot be determined without further information. By measurement its indices can be shown to be $(51\bar{6}1)$.

To find the axial ratios, let us consider a vertical plane through m_1, r_1 and the C axis. If MC represents the trace of r_1 $(10\bar{1}1)$, OM the trace of a horizontal plane and OC the vertical axis (fig. 73), $OM = OC \tan m_1r_1 = c \tan 38° \, 13'$.

But $OM = a \cos 30°$, so that $\dfrac{a}{c} = \dfrac{\tan 38° \, 13'}{\cos 30°} = 0 \cdot 9092$.

CHAPTER III

THE INTERNAL STRUCTURE
OF CRYSTALS

It is difficult to imagine how a crystal growing, let us say, from a saturated solution could take on an orderly geometrical external shape, and, moreover, maintain this geometrical form in all stages of its growth, unless its outward form were merely the visible sign of an orderly homogeneous structure existing throughout the crystal. So that as early as 1784 we find one of the greatest pioneers of crystallography, the Abbé Haüy, putting forward the hypothesis that a crystal is built up of small, identical bricks, each brick, or "molécule intégrante", being an ultimate particle of matter and having a shape related geometrically to that of the complete crystal. A cube, for example, could be built of cubic bricks, a rhombic crystal of rectangular bricks of three different dimensions in length, breadth and height. A hexagonal crystal could be built up by bricks resembling the close-packed cells of a honeycomb; for monoclinic and triclinic crystals the bricks would no longer be rectangular, but would still be identical one with another and regular in their close-packed array.

For many years now the impossibility of this hypothesis has been evident, for modern work in chemistry and physics has supplied undoubted evidence of an open rather than a close-packed structure in all solid matter. The atom itself resembles a solar system in which the central body and attendant planets take up only a minute fraction of the space recognised as belonging to the whole atom, and since this space is, roughly speaking, spherical, close-packing of atoms to exclude all space between them is impossible. Yet the regular arrangement of particles in a crystal is undeniable, and in explanation of this it has become usual to consider these particles as placed at the corners of a regular three-dimensional lattice-work. If adjacent particles are joined by lines—and these lines can conveniently be taken as indicating the existence of certain forces holding the particles in place—a series of similar skeleton cells is formed, which can be taken as the modern equivalent of Haüy's "molécules intégrantes". Each particle may be a single atom or a group of atoms, but if the latter, the arrangement inside a group is invariable throughout the crystal. If the groups are arranged on a cubic lattice-work, each group will be equidistant from the neighbouring group to right or left, above or below, behind or before. It follows that each constituent particle of a group will be at this same distance from the corresponding particle in neighbouring groups; in other words, if the groups of particles form a regular space lattice, each constituent particle of a group will form a similar space lattice when joined to the similar particles in other groups. In general the particles or atoms of a crystal form a series of inter-lacing space lattices, the symmetry of which determines the symmetry of the complete crystal.

It is obvious that a space lattice will possess symmetry just as a crystal possesses symmetry, and it can show planes or axes, simple or alternating. More recently it has been shown that other types of symmetry are possible in these "point systems" that are formed by the particles of a crystal, and *screw axes* and *glide planes* are recognised in addition to simple axes and planes of symmetry. Inform-

ation on such matters must be sought in more comprehensive books; the point to be established here is that in a
crystal identical groups of particles situated at the corners
of a regular space lattice form a system of points which
bear towards one another definite elements of symmetry,
and it is by virtue of this symmetry reflected in its external
form that a crystal is relegated to one or other of the seven
crystal systems. The particles making up a group may or
may not show further symmetry among themselves, and
upon this depends the class within the system to which the
crystal belongs.

It is not yet entirely clear what forces hold the atoms in
their places. In many cases there must be attraction due
to opposite charges on dissimilar atoms, but in some cases,
as for example in crystals of a pure metal where every atom
is identical, a satisfactory explanation of the forces which
pull the atoms into place and hold them there against
considerable attack from without is still to be sought.

There are fourteen possible types of space lattice; that is
to say, points may be arranged in a homogeneous, definite
three-dimensional order according to fourteen different
plans (fig. 74). In six of these the cell outlined by the lattice
is of the same shape as the simple crystals of the various
crystal systems, and a seventh is the rhombohedron. Two
others are the prisms formed by {110} faces in the rhombic
and monoclinic systems, closed by the basal {001}, four
others are of the "centred" type, i.e. the various prismatic
cells with an extra point at the centre of the cell, and the
last is the cubic cell with an extra point at the centre of
each face, or "face-centred cube". The selection of these
types is not arbritary; their determination is based upon
geometrical grounds alone, and they are. the only possible

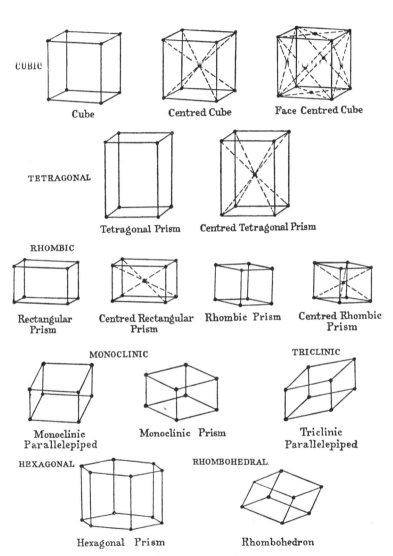

CUBIC

Cube   Centred Cube   Face Centred Cube

TETRAGONAL

Tetragonal Prism   Centred Tetragonal Prism

RHOMBIC

Rectangular Prism   Centred Rectangular Prism   Rhombic Prism   Centred Rhombic Prism

MONOCLINIC   TRICLINIC

Monoclinic Parallelepiped   Monoclinic Prism   Triclinic Parallelepiped

HEXAGONAL   RHOMBOHEDRAL

Hexagonal Prism   Rhombohedron

Fig. 74. The fourteen types of space lattice.

plans upon which regular point systems can be built up. They fall naturally into the seven crystal systems, and were so assigned by Bravais, who in 1848 was the first to give the space lattice theory a definite and mathematical form. The theory provides satisfactorily for the open structure of solid matter, for the homogeneity throughout a substance (since one point in the space lattice bears exactly the same relationship to its neighbours as any other point), and for

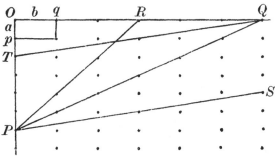

Fig. 75. Two-dimensional space lattice.

the external regularity of crystalline form. Moreover it gives a basis for the law of rational indices, the constancy of crystal angles, and many of the physical and some of the chemical properties of a crystalline solid.

It is difficult to represent a point system in three dimensions; fig. 75 includes two dimensions only, and if the points were linked together by two sets of parallel lines we should have a two-dimensional lattice. Supposing the lattice to be extended in a third dimension perpendicular to the paper, each line of points on the paper would represent the trace of a perpendicular plane of points. Every plane of points in the three-dimensional point system is a possible face in the complete crystal. Supposing *OP* and *OQ* to represent the *A* and *B* axes of a rectangular crystal, the points in the plane of the paper would be the face (001). The line of points *OP* would be the trace of the face (010) or (0$\bar{1}$0), and *OQ* the trace of (100) or ($\bar{1}$00). Two dimensions of the unit cell which is the foundation of the space lattice are shown by the small rectangle drawn at *O*; it should be fairly obvious that a wise and significant choice of parameters would be $a = Op$, and $b = Oq$, so that pq, or *PQ*, would represent the trace of the plane (110). Similarly *PR* would represent the trace of the plane (120)*, and *PS*, or *TQ*, the trace of the plane (310). The basis for the law of rational indices now becomes obvious, and a definite physical meaning can be attached to the parameters of a crystal.

Each of the planes contains a certain number of points per unit of area, some more, some less. *OP* and *OQ*, traces of (010) and (100), contain more points per unit length than *PQ*, or indeed than any other line of points. Hence planes (100) and (010) have the greatest number of points per unit area, and it is obvious that as the indices of a face increase the density of points on the plane representing that face decreases. The simpler the indices of a face, then, the better defined is that face from the point of view of internal structure, so that there is reason for the earlier emphasis laid upon pinacoidal faces and the simpler prismatic faces such as {110}.

The practical importance of certain faces in any individual crystal depends also upon the relationship between

* It must be remembered that intercepts of (120) are $\frac{a}{1}$ on the *A* axis (*OP*) and $\frac{b}{2}$ on the *B* axis (*OQ*).

linkages in the plane of the face and in a direction perpendicular to it, and this varies with the relative spacing of atoms in different directions and with the relationship between the various sorts of atoms which compose the crystal. A wider spacing usually denotes a weaker linkage, but the strength of bonds between atoms depends in the main upon the atoms themselves. If these are all of the same sort the arrangement of atoms tends to be simple, and the forms {100}, {111}, {110} are predominant. It may be remembered that several of the more important elements crystallise in the cubic system, including gold, silver, copper, mercury, and diamond. When the composition is more complex the arrangement of various types of atoms may tend to accentuate certain planes more than others. For example, in the structure of mica certain planes parallel to (001) contain groups of silicon and oxygen atoms, and these alternate with planes containing atoms of magnesium, calcium, etc. The attractive forces between atoms of the silicon layer is much greater than that between atoms of this layer and the magnesium and calcium atoms above and below them; in fact, the linkage in this direction is weak, with the result that the (001) plane in mica grows large in comparison with prismatic planes, but is easily separated from the layers lying above and below. Hence mica shows large basal planes and adopts a tabular habit, while it cleaves easily into thin sheets parallel to the basal plane. This possible difference of linkage along the different directions of a space lattice is the cause of the directional variation in the so-called vectorial properties of a crystal (see p. 4); this point will be taken up in the following chapter.

Experimental proof confirming the space lattice theory

has only been obtained during the last twenty years, as a result of the use of X-rays in connection with crystals. For further detail the student is referred to larger modern text-books of crystallography and especially to the writings of Sir William and Professor W. L. Bragg in England and of Professor Laue of Zürich. Their pioneer work, together with the subsequent work of corroboration and expansion by very many investigators, has not only given full support to the theory but has succeeded in describing and actually measuring the atomic structure in certain crystals, the number of which is almost doubled every year. It has thrown light upon the problems of chemical combination and upon many other matters not strictly connected with the study of mineralogy, as well as causing the mineralogist to revise his opinion in some particulars. For example, not so many years ago minerals were divided into a large class of crystalline substances and a small class of non-crystalline or amorphous minerals which showed no crystal outlines and even under the microscope gave no indication of crystallinity. These were regarded as solids in which, by reason of the quickness of their formation or other causes, the constituent atoms had been unable to settle down side by side, rank by rank, into an orderly array. X-rays provide a more searching means of examination, however, and their use has shown that few, if any, solid substances are entirely lacking in crystalline structure. In some the solid may consist of sub-microscopic crystalline blocks lying up against one another higgledy-piggledy; each block consti-tutes a perfect crystal on a minute scale, but even under the microscope the lack of orientation between the blocks annuls any orderly, crystalline effect of the whole solid. Even substances of vegetable origin have yielded evidence of crystalline structure, and it appears that far from being the ideal type of solidification, attained only under certain narrow conditions of physical environment, the crystalline structure is very widespread, and that atoms collecting themselves together into simple or compound substances show a decided preference for order and regularity.

CHAPTER IV

GENERAL PHYSICAL PROPERTIES
OF CRYSTALS

TWINNING

Many of the characteristics of a crystal are dependent upon the regularity of its internal structure, and this shows itself in nothing more clearly than in the phenomena dealt with under this heading.

It has already been stated that a mineral fragment does not always consist of a single crystal; groups of crystals of various sizes, ultra-microscopic, small, or large enough to be easily recognisable by the naked eye, are of very common occurrence, and the individual members of such a group generally seem to have been thrown together haphazard. But in some cases we find two crystals or a group of crystals joined together in a fashion so regular and so intimately related to the structure of each individual that obviously some cause more orderly than chance has been at work. Crystals of this kind are termed *twinned crystals*.

Twinned crystals are not all of one type. In some cases one could break the individuals apart, and each would form a simple crystal complete as far as the plane of fracture; these are known as *contact twins*. In other cases it appears as if two crystals had grown into and through each other so completely that separation would be impossible; this is what is known as *interpenetration twinning*. In a third type a whole series of crystals are united each to its neighbours in a regular fashion, and this is termed *repeated* or *polysynthetic twinning*. Repeated twins often form what is apparently a simple crystal, but the twinning is indicated by the banded appearance of certain crystal faces, which, like a new-mown lawn, show alternate bands of dark and light. In all cases the twins are related one to the other in such a way that one individual can be transposed into the position of its twin either by reflection across a plane, known as the *twin plane*, or by rotation through 180° about an axis known as the *twin axis*; the twin plane is always parallel to a possible face of the crystal, and the twin axis is always perpendicular to a possible face or parallel to a possible crystal edge. Many twins can be regarded as either reflection or rotation twins at will, but in some cases no twin plane exists and in others there is no twin axis. If the two parts are separable, the plane in which they join is known as the *composition plane*; this again is always parallel to a possible face in the crystal, and may coincide with the twin plane.

A few examples may make these terms clearer.

Fluor spar often shows interpenetration twinning, two cubes combining as in fig. 76. If one part, the shaded part for example, is rotated 180° about the triad axis indicated, i.e. about the normal to the octahedral faces $(11\bar{1})$ and $(\bar{1}\bar{1}1)$, it will come into a position parallel to that of the unshaded part. Alternatively, the shaded cube would give in a mirror held parallel to $(\bar{1}\bar{1}1)$ a reflection oriented parallel to the unshaded twin. In this case the crystal is said to be twinned on, or parallel to, an octahedral face.

The plagioclase felspars are characterised by their repeated twinning. In the type known as *albite twinning* the alternating layers show most clearly on the basal planes, the twin plane being (010), but the twinned crystal preserves the shape of a single crystal (fig. 77).

Calcite twins in various ways; fig. 78 shows the characteristic re-entrant angles on a crystal whose twin plane and composition plane is (0001) and whose twin axis is the vertical triad axis perpendicular to (0001).

Fig. 79 shows a triplet, or elbow-twin, of rutile (TiO_2), the twin plane being the plane face (101).

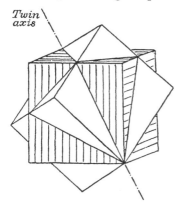

Fig. 76. Interpenetration twins of   Fig. 77. Repeated twinning
            fluor spar.   in albite.

To regard a twinned crystal as composed of two separate crystals which, starting from separate centres of crystallisation, have grown side by side, finally meeting and joining together in the composition plane, is to misconceive the fundamental nature of twinning. In the same way the classification of twins as twins due to rotation or to reflection is purely utilitarian and has no relation to the actual processes of their formation. A twinned crystal is essentially a single crystal which has developed by growth about a single nucleus. It differs from a single crystal in that at a certain stage of growth the particles about to link themselves to one side of the crystal did so in a reversed position which was adhered to throughout the growth of the second twin. In repeated twinning this reversal of orientation takes place over and over again along parallel

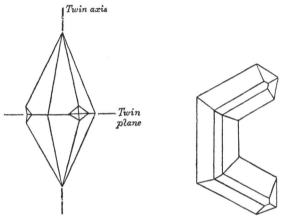

Fig. 78. Calcite twin.   Fig. 79. Elbow-twin
                                        of rutile.

planes; in interpenetration twins the part common to both twins is built up partly on the original lattice system and partly on that system in the reversed position.

The twinning of gypsum affords a good example of this reversal process. Gypsum ($CaSO_4$) forms simple mono-clinic crystals, and twins with (100) as twin plane are common (fig. 80). The fundamental space lattice is a series of planes of particles parallel to ab and ac, running perpendicularly through the paper (fig. 81). We can imagine particles falling into place in these planes on both sides of centre of crystallisation O. At the time when *ac* represented the left-hand limit of the solid crystal, however, for some reason still unknown the orientation of the structure changed, and additional particles ranged themselves in planes parallel not to *ab* but to *ab'*, where *ab* and *ab'* are

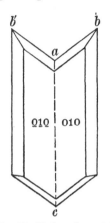

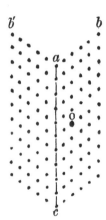

Fig. 80. Twinned crystal   Fig. 81.
of gypsum.

equally inclined to *ac*. The left half of the crystal was therefore built up in such a way that it became the mirror image of the right half, and the composition plane runs through *ac* perpendicular to the paper, being parallel to (100). The crystal can be described as twinned on the face (100), or, alternatively, as twinned about an axis *ac*, which is the zone axis of faces (100), (110), etc., and is also the vertical crystallographic axis.

Twinning, then, is an affair of internal structure, and the positions of twin axis and twin plane are governed by the same laws as determine the position of crystal faces and the sizes of crystal angles.

CLEAVAGE

Reference has already been made to the phenomenon of *cleavage* (see p. 78), and in dealing with the internal structure of crystals a cause was assigned to the readiness shown by mica to split parallel to the basal pinacoid into sheets of a thinness limited only by the skill of the operator and the delicacy of his tools.

Many other minerals exhibit a similar tendency, the cleavage planes being always parallel to a face, or a possible face, of the crystal. The ease of splitting varies, as does the perfection of the plane surfaces created by the split, and accordingly cleavage is said to be *perfect, good, fair, imperfect*, etc., but even imperfect cleavage is easily distinguishable from the rough, uneven surface created when a mineral is broken in any other direction. If a mineral possesses cleavage it can be split, not only here and there within the crystal, but at any point whatever as long as the splitting is allowed to take place parallel to a certain face.

Since it is purely the result of atomic arrangement, cleavage falls into line with the other indications of crystal symmetry. For example, calcite, crystallising in the rhombohedral system, has perfect cleavage parallel to one pair of rhombohedral faces, and this is accompanied by two other sets of exactly similar cleavage parallel to the other two pairs of rhombohedral faces; the unique triad axis of symmetry is indicated by the arrangement of similar cleavage planes as well as by the arrangement of similar crystal faces. Mica has perfect basal cleavage, and in this case the symmetry does not demand any second set of cleavage. Other examples of good cleavage are to be found in rock salt, which possesses perfect cubic cleavage parallel to three perpendicular planes, and in diamond, whose perfect octahedral cleavage is a boon to the cutters, giving them four directions in which bright, smooth facets can be produced with ease. In the rhombic system, since no two pairs of pinacoids are similar, one cleavage parallel to one pinacoid can exist alone. If the cleavage is prismatic, however, the symmetry of the crystal demands two sets of cleavage planes equally inclined to a pinacoidal face. This type of cleavage is exhibited by barytes or heavy spar; two types of cleavage can occur in the same mineral, and barytes also shows perfect basal cleavage. Prismatic cleavage is characteristic also of the pyroxenes and amphiboles, while the felspars show perfect basal cleavage and a less perfect cleavage parallel to the pinacoid (010).

The tendency to cleave is sometimes indicated by a series of fine striae visible on faces perpendicular or nearly perpendicular to the cleavage planes; where there is a perfect cleavage parallel to well-developed faces, as in mica or calcite, these faces may show a stepped formation where fragments have broken off along the cleavage planes. Under the microscope cleavage planes appear as dark parallel lines of varying persistency running across crystal sections. The type of cleavage and the angle between two sets of cleavage are most useful indications of the identity of the mineral, and in fragments which have lost their crystal outlines cleavage planes or traces of planes can be used to show the orientation of the faces that have perished.

For example, the minerals hornblende and augite are very similar in a hand specimen. They each possess perfect cleavage parallel to (110), but since the axial ratio $a : b$ is

1·092 in augite and 0·551 in hornblende, the angle between the two sets of cleavages differs considerably in the two minerals, being 93° in augite and 56° in hornblende. Figs. 82 and 83 show typical sections of augite and hornblende; it will be noted that the cleavage of augite is rather

Fig. 82. Prismatic cleavage
in augite.

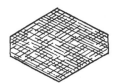

Fig. 83. Prismatic cleavage
in hornblende.

more interrupted and less perfect and persistent than that of the amphibole.

Again, the felspars possess perfect cleavage parallel to (001) and less perfect cleavage parallel to (010). In the monoclinic crystals of orthoclase the sets of cleavage traces are at right angles to each other in sections parallel to (100), while in the triclinic plagioclases the angle between them is less than 90°, and its determination is one method of identifying the various members of the plagioclase family.

FRACTURE

Some minerals possess no tendency to cleave; quartz, for instance, breaks in an irregular way, showing a fractured surface resembling that of broken glass. The reason for this is evident from the study of a space lattice model of the quartz crystal, for the characteristic feature of the model is the spiral arrangement of the atoms within a hexagonal framework, and no well-defined planes exist within the structure at all. Moreover, minerals with cleavage in some directions fracture in other directions in an irregular fashion dependent upon their atomic arrangement. Fracture is termed *smooth* or *even* when the fractured surfaces are approximately plane, though full of small irregularities, *uneven* when the surface is quite irregular, *hackly* or *splintery* when the surface is jagged. The characteristic series of curves resembling those of a shell which are visible on the broken surface of glass, flint or quartz have given rise to the descriptive term *conchoidal* in relation to fracture of this type.

GLIDE PLANES AND PARTING

Some minerals, generally those of a relatively soft nature, exhibit the capacity of developing *glide planes*. These resemble cleavage planes in their relationship to the symmetry of the crystal and often also in their appearance under the microscope, but they differ from true cleavage planes in that they only appear at places where there has been definite molecular disturbance in the crystal, the molecules on one side of the plane gliding over those on the other side. In many cases the glide planes are parallel to the natural cleavage of the mineral, but this is not invariable; in rock salt, for example, the cleavage is cubic, but glide planes can be induced by suitable pressure in six directions parallel to the dodecahedral faces. They often constitute planes of weakness in the crystal and so are sometimes termed *planes of parting*.

The most striking case of molecular gliding is that of calcite. The pressure of a blunt knife-edge on one of the primary rhombohedral edges may in skilful hands produce a slipping of the molecules on one side of the knife-edge, so that, layer after layer, they glide sideways from the knife into positions the reverse of those in the unaffected part of the crystal, forming what is known as a *secondary twin*. The twin plane is the secondary rhombohedron which truncates the edges of the primary rhomb.

COLOUR

Colour is generally the first property of a mineral to strike the eye, but unfortunately it is not one upon which great reliance can be placed for purposes of identification; it may be an essential property of the mineral or it may be merely lent by extraneous agents. Many minerals appear black and opaque by reflected light, but a very thin slice prepared for microscopic examination is often found to be transparent and by transmitted light appears green, yellow, brown or red. Mineral ores in general remain opaque even in thin slices, but by reflected light they may have their distinctive colours, grey, red, green, silver or brassy. It is unsafe to rely upon an exposed face to show the true colour of the mineral, for a film of tarnish gives a very different effect, and some minerals which oxidise readily on exposure to the air, such as the sulphide of iron, marcasite, only show their real colour on a freshly fractured surface.

Some minerals possess distinctive colours which are sufficiently constant to be used for purposes of identification, but in many cases coloration is due to the presence of impurities and may vary greatly in different specimens of the same mineral. For example, quartz may be found colourless or smoky grey, or it may show the yellow of the cairngorm or citrine, the purple of the amethyst or the pink of the so-called rose-quartz. Ruby and sapphire are differently coloured forms of the usually grey mineral corundum (Al_2O_3); the "Blue-John" of Derbyshire is in-distinguishable chemically from the green and colourless varieties of the same mineral fluor spar (CaF_2). Again, a mineral may acquire a colour due to peculiarities in its internal structure. The play of colours in an opal is due to the variation of refractive index in the very thin layers of which the mineral is built up; in other cases colour may be due to the presence of minute cracks in the crystal in which thin films of air produce the so-called interference tints that are also seen when a drop of oil spreads in a thin film over the surface of a puddle. Occasionally very thin plates or flakes of a different composition are included in a parallel fashion throughout a mineral, and in the same way these produce an iridescent effect; the felspar labradorite affords a striking example of borrowed colour of this sort. The same kind of effect is also produced by chemical change taking place along parallel planes of weakness in a crystal, a process which is known as *schillerisation*. A certain coarse-grained rock known as lauvikite containing large crystals of a schillerised felspar is in great favour as a polished ornamental stone for certain types of building.

STREAK

The colour of a powdered mineral is sometimes different from that of the solid crystal. Most non-metallic minerals yield a colourless powder, but the characteristic colours of some metallic minerals when crushed afford a useful clue to their identity. For instance, the dark purplish oxide of iron, haematite, yields a dull red powder, the brown hydrated oxide of iron, limonite, gives a characteristic yellow powder, while the powder of the brassy sulphide, pyrites, is almost black.

The characteristic colour is best exhibited by grinding the mineral fine and then pressing the powder on to a white surface; this produces a mark which is technically known as the *streak* of the mineral. The easiest way to obtain the streak is to rub a corner of the mineral on a hard, unglazed piece of porcelain or on the fractured surface of light-coloured flint or chert.

LUSTRE

The surfaces of a mineral examined by reflected light differ considerably from one another in appearance, some resembling polished metal, others in extreme contrast looking dull and greasy. The effect is governed partly by the refractive index of the mineral and partly by internal structure. The very high lustre shown by opaque minerals such as iron pyrites and many other mineral ores is termed *metallic*; a few translucent, deeply coloured minerals, including the black variety of mica, show an imperfect metallic or *sub-metallic* lustre. Most translucent minerals possess a *vitreous* lustre, the lustre of broken glass; those of very high refractive index, for example diamond and cassiterite, or tin-stone, possess a higher lustre described as *adamantine*, and this too is the property of some minerals of high specific gravity. In contrast to these, a few minerals, notably nepheline, are distinctly *greasy* in appearance; their lack of lustre seems to be the result of their comparative softness. A *pearly* lustre is exhibited by some mineral surfaces as a result of cleavage parallel to that surface; in orthoclase, for example, the perfect cleavage parallel to the basal face (001) gives a pearly lustre to that face, while the other faces show a higher, vitreous lustre. Minerals of a fibrous structure may exhibit a *silky* lustre; this is well seen in the case of the aptly named variety of gypsum, satin-spar.

HARDNESS

The hardness of a mineral is a very important property from the commercial point of view as well as being of great use for identification purposes. The wide use of talc (a member of the amphibole family, and not to be confused with the so-called talc used for lamp chimneys, windows, etc.) is due largely to the unrivalled softness of the mineral; on the other hand, no mineral can be used as a gemstone, however suitable are its other properties for the purpose, unless it is hard enough to withstand a certain amount of abrasion. The diamond owes its value partly to its very high refractive index and dispersive power, but still more to its extreme hardness, and quartz, often in the shape of fine sand, is an integral component of many abrasive and cleansing preparations because of its power to scour away softer substances.

Hardness can be defined as the degree of difficulty offered by a substance to the separation of its ultimate particles; it is thus directly dependent upon the intimate structure of the substance, and since in crystals this varies in different directions, the hardness of a mineral is a vectorial property. The ordinary method of measuring hardness is by attempting to scratch the face of a mineral with a file or knife or with a sharp edge of another mineral; if the scratcher is harder than the mineral, it tears apart the particles along the line of scratch, leaving a tiny groove which may need a lens for its detection. The harder substance may appear to have been damaged by reason of a powder left on its surface by the abraded particles of the softer mineral, but this can be wiped away and the surface is seen to be left untouched. For this reason the definite production of a groove and not merely of powder must be taken as evidence of scratching.

Minerals show a wide range of hardness. The degree of hardness is described by reference to a scale of ten minerals drawn up by Friedrich Mohs in 1820, commonly known now as Mohs' Scale. Each mineral is harder than the one immediately below it on the scale and is denoted by a number; the whole scale is as follows:

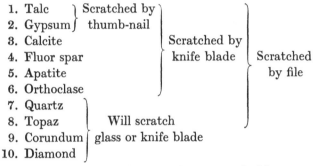

Quartz will scratch felspar but is itself scratched by topaz; diamond is the hardest mineral known and talc the softest. The thumb-nail and a penknife give a rough and ready means of deciding the hardness of very many minerals, as is indicated at the side of the scale, and furthermore the ease with which the mineral is scratched by either helps to differentiate between, say, minerals of hardness 4 and 6. Quartz and a knife blade are so nearly alike in hardness that they may both scratch each other simultaneously.

A more accurate method of determining hardness is by means of a *sclerometer**. The main principle of the method is the abrasion of a mineral by a hard point, usually diamond,

* See especially Tutton, *Crystallography and Practical Crystal Measurement*.

which bears on it with a measured load and works to and fro along a definite line. The amount of abrasion is measured either by the minimum load necessary to produce a scratch, or by the weight of material abraded by a standard load in a given number of traverses. If the hardness of a crystal face is thus measured in different directions and the results plotted as distances from a centre in these directions, a closed curve is obtained which is not circular but which portrays the symmetry of the crystal with regard to that face. For the (100) face of a cubic crystal this *curve of hardness* is a square with hollowed sides and rounded corners; for a rhombohedral face of calcite it is a symmetrical three-lobed curve.

SPECIFIC GRAVITY

There is a large range in the specific gravity of minerals, from values little above unity for some of the so-called non-metallic minerals to values nearly twenty times as great for the heavy metals. Gold has a specific gravity of 19, iridium of 23. Almost all the common minerals, how-ever, have specific gravities between 2 and 7, and the most familiar rock-forming silicates seldom possess a specific gravity above 4.

The density of a mineral depends upon two factors, the density of the atoms of which the mineral is composed, and the way in which those atoms are packed together. Com-pounds of heavy atoms tend to have a high density; for example, compounds of the heavy metals barium and lead are usually characteristically heavy, while compounds of aluminium, unless they also contain heavy metals, are usually of low specific gravity. But even light atoms can form heavy compounds if they are packed together in a sufficiently dense array. The light carbon atom of atomic weight 12 forms the diamond of medium specific gravity 3·5, a value slightly higher than that of garnet which is made up entirely of atoms heavier than carbon. But graphite, with its looser arrangement of carbon atoms, has a specific gravity as low as 2·1. As far as specific gravity is governed by closeness of packing, its values keep parallel with those of the hardness of minerals; this becomes very evident on reference to the tables drawn up by Dana*. There are many exceptions, however, caused by loose packing of heavy atoms or close packing of light atoms. Gold, with its high specific gravity of 19, is a very soft metal, and the carbonate and sulphate of lead, cerussite and anglesite, with high specific gravities, have a hardness as low as 3 to $3\frac{1}{2}$. On the other hand, andalusite (Al_2SiO_5) contains no heavy atoms and has a specific gravity of 3·2, but its particles are so tightly linked that its hardness is as high as $7\frac{1}{2}$. In a group of minerals having the same structure but different chemical composition, such as the plagioclase felspars, the garnet family, or the isomorphous sulphates of barium, strontium and lead, the specific gravity increases as the varying atom increases in weight. As calcium gradually replaces sodium in the plagioclases the specific gravity rises from 2·62 in albite to 2·76 in anorthite; the specific gravity of celestine ($SrSO_4$) is 3·9, of barytes ($BaSO_4$) is 4·3, and of anglesite ($PbSO_4$) is 6·3.

For the geologist's purpose it is usually sufficient to know the specific gravity of a mineral to two places of decimals. Care has to be taken to guard against a diminu-

---

* Dana, *Text-book of Mineralogy*. For further information see also Miers, *Mineralogy*, and for methods of determination Tutton, *Crystallography and Practical Crystal Measurement*.

tion in value caused by inclusions or air-bubbles, and especially in the case of fibrous crystals or other types of aggregates it is advisable to reduce the crystals to small fragments before making a determination. The usual methods are employed, modified to some extent to suit the special needs of the geologist, and various instruments have been designed to provide him with a ready means of estimating specific gravity to the accuracy that he requires. Various types of balance may be found described in the standard text-books for the determination of specific gravity by the usual hydrostatic methods, that is, by the comparison of the weight of a fragment with the weight of the water that it displaces, or, in other words, with its loss of weight in water.

A second method makes use of the specific gravity bottle or *pycnometer*, a small stoppered bottle of known weight P. The mineral, reduced to small fragments, is placed in the bottle, and the combined weight of the bottle and mineral is found (W_1); the bottle is then filled with distilled water, boiled to get rid of air-bubbles, cooled, filled up with boiled water and weighed again (W_2). Finally, the pycnometer is emptied, filled again with distilled water and weighed once more (W_3). The weight of the mineral fragments in air is then $W_1 - P$, and the weight of the water that they displace is $(W_3 - P) - (W_2 - W_1)$, so that the specific gravity of the mineral is given by $\dfrac{W_1 - P}{(W_3 - P) - (W_2 - W_1)}$. If necessary, a correction can be made for variations in temperature, the specific gravity of a substance being defined as its density relative to that of water at 4° C.

A third method involves the use of heavy liquids. A solid sinks in a liquid of lower density and floats in one of higher density than itself; in a liquid of equal density it remains suspended at any level to which it is thrust. By using a heavy liquid such as methylene iodide (sp. gr. 3·3) and diluting it gradually with benzene (sp. gr. 0·98), a mixture may be obtained in which a certain mineral particle neither floats nor sinks. The specific gravity of the liquid is then equal to that of the mineral and can be determined by means of a pycnometer. Various liquids are used for this purpose; Klein's solution, an aqueous solution of cadmium borotungstate (sp. gr. 3·28), can be used except with carbonates, with which it reacts; others in general use are aqueous solutions of potassium mercuric iodide or of barium mercuric iodide, and also bromoform (sp. gr. 2·88).

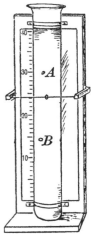

Fig. 84. Sollas diffusion column.

An outcome of this method is the *Sollas diffusion column* (fig. 84). If a test-tube is partially filled with methylene iodide upon which is poured an equal quantity of benzene, the two liquids will gradually diffuse into each other, forming after some hours a column whose specific gravity varies continuously and regularly from that of nearly pure methylene iodide at the bottom to that of nearly pure benzene at the top. The column can be calibrated by introducing into it two fragments of known specific gravities; each will remain suspended at that level where the specific gravity of the liquid column is equal to its own. A scale is set up by the side of the column, and a movable wire is stretched horizontally across the tube and reads on the scale. A mirror is fixed behind the tube so that the reading may be taken when the wire coincides with its own image, thus avoiding errors due to parallax. The position of any mineral fragment can then be read on the scale and compared with the positions of the standard fragments, and the specific gravity of the specimen can be deduced. For if the specific gravities of the standard fragments A and B are G_A and G_B respectively, and their readings on the scale P_A and P_B, then the specific gravity G of a mineral taking up an intermediate position which gives a reading P on the scale is given by the equation

$$\frac{G_A - G}{G - G_B} = \frac{P_A - P}{P - P_B}.$$

CHAPTER V

OPTICAL PROPERTIES OF MINERALS

INTRODUCTION

We have seen already that minerals are characterised by various external physical and morphological properties, many or all of which are the result of internal structure. There is no possibility at present of magnifying that minute internal structure so that it actually becomes visible; we can only infer its nature from external evidence. There is a second line of attack, however, and that is to examine the influence of the internal structure on something that has passed through the crystal, and in so doing has been affected by the regular array of particles of which the crystal is composed. The most searching enquiry is carried out by passing X-rays through the crystal, but this is not at present within the power of the ordinary student. Another and somewhat similar investigation can be made by means of a beam of light, which is also affected in various ways by its passage through a crystal; but before these effects can be described it is necessary to recall the general characteristics of the phenomenon known as light.

Until comparatively recently it has been held that the sensation of light is due to a wave-motion in which the energy is handed on from particle to particle until finally it excites the nerves of the retina to produce in the brain the sensation of light. Since light can travel through empty space even more easily than through a medium such as air, water or a transparent substance, and since no method is known by which energy can be transported without the presence of a medium, it becomes necessary to postulate the existence of an entirely hypothetical substance, the *ether*, which pervades all space, whether that space is what we call empty or filled with matter in any form. The energy of the disturbance called light is supposed to pass from particle to particle of this weightless, invisible ether, but the disturbance is at the same time affected by the nature of the material particles which lie in its path.

Recent advances in theoretical physics make it appear doubtful if light is really of such a nature as this; there is evidence that in some respects light behaves less like a wave-motion and more like the train of minute, swiftly moving corpuscles which was Newton's conception of a light ray. Moreover, it has so far proved impossible to justify the hypothetical existence of the ether of space. But the theoretical explanation of the optical properties of crystals is still usually based on the wave-theory of light, and on this basis all the phenomena observed are explainable in a general way. The hypothesis of a simple system of waves passing through an orderly arrangement of particles does not, however, satisfactorily explain all details of the most complex phenomena observed, and when this occurs, it must be remembered that the wave-theory of light, with its indispensable handmaiden, the all-pervading ether, is but a hypothesis and probably only an approximation to the truth.

The wave-theory of light will be assumed to be true, however, for the purposes of the following elementary description of the optical properties of crystals. The vibrations constituting the wave-motion are treated as if they were purely mechanical; in reality they are electro-magnetic in character, and the to-and-fro movement is of electric and magnetic charges rather than of material particles.

THE NATURE OF LIGHT

For the present, then, light is assumed to be a disturbance of the ether propagated in straight lines by a movement of particles to and fro in directions perpendicular to the direction of propagation of the light. The disturbance takes the form of a succession of waves, similar to those caused by dropping a stone into water, the particles concerned moving up and down through a mean position of rest, while the disturbance as a whole spreads outwards from the source in a direction perpendicular to the vibration of the individual particles. The disturbance is generated by the particles of a body being set in regular vibration; this causes the ether particles next to the body to take up the vibration—that is to say, as any particle vibrates it pulls the adjacent particle out of its position of rest and causes it to perform a like vibration. Hence the energy of vibration is handed on through an infinite number of particles, each particle lagging slightly behind the particle from which it obtained the energy of vibration.

Fig. 85 a shows the displacements at any one moment of the particles along the line of a ray of light emanating from a point source at A. It is only possible in a diagram to show displacement in the plane of the paper, but it must be remembered that ordinarily the particles may be displaced in every direction perpendicular to the path of the ray.

Each particle along the line of the ray AA' moves up and down between the two dotted lines BB' and CC' with simple harmonic motion in much the same way as the bob of a pendulum swings from side to side through its position of rest. The velocity and acceleration at each point of its path between B and C are those of the projection P on BC

of a point P' which travels round the circle $A'B'C'$ with
constant speed (fig. 85 b); that is to say, the velocity is
greatest at A, decreases until the particle comes to rest
at B, where motion starts in the backward direction,
quickening up as far as A and slowing down again until
the particle comes momentarily to rest at C before moving

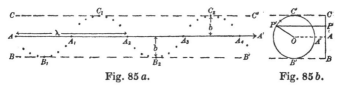

Fig. 85 a.   Fig. 85 b.

in the same way back to A. Then the whole cycle of
movement is repeated.

The particle A_2, for instance, has vibrated through its
original position a certain number of times, and is about to
move up to line CC' again; particle A_4 is in the same state
but has performed the whole vibratory movement once less
than A_2. Particles A_2 and A_4 are therefore said to be in the
same phase, or to have a phase difference of one wave-
length. Particle A_1 is also momentarily in its zero position,
but it is about to move down to line BB', being drawn
downwards by its neighbour on the A side which is already
slightly displaced towards BB'. Hence the movement of
A_1 is exactly opposite to that of A_2 and A_4, and A_1 and A_2
are said to be in opposite phase, or to have a phase difference
of half a wave-length.

Particles B_1, B_2, C_1 and C_2 are in extreme positions;
particles between B_1 and A_1, A_1 and C_1, etc., are in intermediate positions, all moving perpendicular to AA', which
is the direction of propagation of the ray. The particles are
obviously arranged to form a regular series of troughs and

crests which travel onwards towards the right, while the particles themselves only move in an up-and-down direction.

The section of the wave between A and A_2, two particles in the same phase, defines the wave-motion; the rest of the system of waves is made up of repetitions of this section. The distance between A and A_2, or between any particle and the nearest particle to it in the same phase, is therefore known as the *wave-length*, and is denoted by the Greek letter λ. The distance between the lines BB' and CC' defines the breadth of the wave-motion, and the distance of either of these lines from the line AA' is known as the *amplitude* of the wave. An increase in the energy of the vibration set up by the source at A would have the result of swinging the particles further from their position of rest; hence increased energy produces an increase in amplitude, and this makes itself known to us as an increase in the intensity of the light perceived. It can be shown that the intensity of the light varies as the square of the amplitude.

The time taken for a particle to perform a complete vibration is known as the *period* of the wave-motion; this is constant for any one set of rays, since it is a function not of the medium through which the light travels but of the source of vibrations itself. The number of vibrations sent out by the source per unit of time is known as the *frequency*; the frequency is therefore the reciprocal of the period*.

---

* The equation of the curve in fig. 85 a is

$$\eta = b \cos 2\pi \left(\frac{t}{\tau} - \frac{x}{\lambda} \right),$$

where η is the displacement of any particle P from its position of rest P_0, b is the amplitude AB, τ the period of the vibration, λ the wave-length, x the distance of P_0 from the source A and t the time the disturbance has taken to travel from A to P_0.

If the frequency is n, and the wave-length is λ, the waves must travel with a velocity $n\lambda$. Hence the velocity of the wave is directly proportional to the wave-length so long as the frequency remains unchanged. As a wave travels through different media the different types and arrangements of material particles affect the wave-motion, and both the velocity and the wave-length may change as the wave passes from one medium to another, but so long as the vibration of the source remains unchanged the velocity and the wave-length always remain in the same proportion to one another.

The same source may vibrate in such a way that trains of waves of different lengths are given out simultaneously. Light waves travel through empty space with a constant velocity of 186,000 miles per second, whatever their wave-length. In material media, however, the velocity varies with the wave-length, and short waves travel more slowly than longer waves. When the waves impinge on the retina the effect produced also varies with the wave-length, and as this increases from 380×10^{-7} cm. to 760×10^{-7} cm. we see successively all the colours of the rainbow from the short violet waves to the red rays which have a wave-length twice as long. When we receive the sensation of white light our eyes are really receiving light of these and all intermediate wave-lengths, but we are not able to sort out the different component vibrations without some mechanical device, such as a glass prism. Light is generally described in terms of its wave-length, but since the wave-length varies with the medium through which the light is travelling, it would be more proper to describe it in terms of its frequency which is independent of the medium. This would, moreover, give a parallel classification to that of sound waves whose pitch is always defined in terms of frequency. However, the classification of colour by its wave-length *in vacuo* is now an established convention.

The sun emits rays of very varied frequencies, and the total effect of all its light rays upon our eyes is what we term " white light ". It is not necessary, however, to include all these frequencies to produce the same effect artificially; a mixture of three or even two colours in the right proportions looks as satisfactorily white to us as does the sunlight. Blue and yellow, green and magenta, or blue, green and red, if blended in the right proportions, give the appearance of white. An extra amount of one wave-length in a white-producing mixture of this kind will give a corresponding tint to the combined beam; in the same way, if we deplete white light wholly or in part of one of its component colours, the residual light will appear tinted with what is known as the complementary colour to that which has been extracted. For instance, if the blue rays are extracted from white light the resulting light appears yellow. The importance of this will appear later in dealing with the phenomena of interference of rays.

Waves of the same type exist whose wave-lengths are outside the limits which produce on the retina the sensation of light. Waves of slightly longer wave-length than that of red light can be detected and are known as the infra-red rays, while those whose wave-length is anything between 3 yards and 20 miles are familiar to us as so-called wireless waves. At the other extreme beyond the limits of the violet light are the waves known as the ultra-violet, while still shorter wave-lengths produce X-rays, which in turn give place to gamma-rays whose wave-length is of the order of a million-millionth of an inch.

So far we have considered only a single ray whose path through space can be represented by a straight line. In reality a source of light sends out rays in all directions, just as a stone dropped into water sends out disturbances in all directions on the surface of the water, the crests and troughs forming ever-widening circles round the origin of disturbance. In the same way the waves of light, restricted to no one plane, will spread out around the source, and at any one instant the crests and troughs of the waves will form a series of curved surfaces concentric about the source. The surface formed by all particles in the same phase, such as all those of the crest of a wave, is known as the *wave-surface* of the disturbance. It is obvious that if the waves spread evenly in all directions the wave-surface is a sphere, and at every point it is perpendicular to the actual ray of light that has travelled from the source to that point. If, however, as is the case in many crystals, the rays can travel more quickly in one direction than another, the wave-surface will take the form of an ellipsoid, and the rays showing the paths of the vibrations from the source to the surface will no longer always be perpendicular to the wave-surface.

Polarised Light.

In an ordinary ray of light vibrations are taking place in all directions perpendicular to the path of the ray, and the vibrations of any one particle may change in direction from moment to moment. It is possible, however, by certain methods to restrict the direction of vibrations, so that a ray is carried in a horizontal direction, say, by vertical vibrations only and is deprived of all vibrations in a horizontal plane or in planes intermediate between that and the vertical. It is as if the vibrations of ordinary light had been passed through a fine vertical grid which had damped out all vibrations that were not accurately parallel to its narrow slits. The intensity of the rays would have been diminished by passing through the grid, and the vibrations would continue in a vertical direction only unless they were further interfered with. Rays of this type are known as *plane polarised light,* since the vibrations all lie in one plane through the path of the ray. It is possible also to restrict light vibrations so that each particle moves, not in a straight line, but in an ellipse or a circle; in these cases the light is said to be *elliptically* or *circularly polarised.* Light of this type is not of any general use in the examination of crystals and will not be discussed further. Plane polarised light, however, is of very great importance in dealing with the optical properties of crystals, and it is vital to get a clear conception as to its nature.

The passage of light through a crystal may be modified by the arrangement of the crystal particles in its path; it is as if these particles behaved as obstacles set in regular ranks restricting the free vibrations of ordinary light much in the same way as a grid. By studying the way in which the emergent vibrations have been modified, we can get evidence of the arrangement of the obstacles, and the problem is obviously simplified by using light vibrating in one direction only. The production and analysis of polarised light therefore becomes a process of great importance to the petrologist, and petrological microscopes are fitted with the means to make this possible.

There are two planes of obvious importance in a ray of polarised light—the plane containing the directions of the ray and of the vibrations, that is to say, the plane of the paper in fig. 85, and the plane through the ray perpendicular to this, in which there is no movement of particles at all. By a rather unfortunate choice this latter plane was the one chosen by the early investigators as the *plane of polarisation,* so that we have to envisage all vibratory motion as taking place perpendicular to the plane of polarisation. The point is of little importance in reality, for according to the modern electro-magnetic theory of the nature of light, there are vibrations in both planes, a magnetic force acting in the plane of polarisation and an electrical force in what we have taken to be the plane of mechanical vibrations of the ether particles. On the basis of the wave-theory of light it is this plane of vibrations that is of importance in dealing with crystal optics, and reference will be made constantly, not to the plane of polarisation, but to the plane of vibration of the particles along the path of the ray.

Relationship between Ordinary Light and two Perpendicularly Polarised Rays.

It is possible to consider a ray of ordinary unpolarised light as the equivalent of two rays polarised in perpendicular planes but otherwise identical.

Suppose that *POP′* represents the path of a particle vibrating with simple harmonic motion about its position of rest at *O* (fig. 86). The particle may be regarded as affected by a periodic force acting along *OP*. But it may equally well be regarded as affected by two periodic forces acting simultaneously and in phase along the two perpendicular lines *OQ* and *OR*. One force pulls the particle from *O* to an extreme position *Q*, while simultaneously the second force pulls it through a displacement OR; the combined result will be that the particle is displaced to a position P. It can be shown mathematically that any two simple harmonic motions of equal period can be combined in this way to give a resultant simple harmonic motion of the same period.

Suppose, then, that POP' represents a vibration of a ray of ordinary light travelling perpendicularly through the paper. The direction POP' is unrestricted within the plane of the paper, but whatever it may be we can always resolve it into two equivalent vibrations of the same period along QOQ' and ROR'; part of the energy of the POP' vibra-

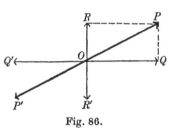

Fig. 86.

tion will be used to give the QOQ' vibration and the rest will form the ROR' vibration. In other words, a ray of ordinary light can always be regarded as equivalent to two rays of the same period polarised in perpendicular planes, or, conversely, two such perpendicularly polarised rays, if travelling along the same path with the same velocity, can be compounded to form a single ray of the same period, polarised in a direction dependent upon the relative phases and intensities of the two component vibrations. Fig. 86 illustrates the case when these vibrations have a phase difference of any number of half wave-lengths, so that each tends to move the particle from its zero position at the same moment. If, however, the phase difference is not a whole number of half wave-lengths the polarisation of the resultant is not plane but elliptical; when the phase difference is any odd multiple of a quarter wave-length, the polarisation becomes circular.

Light travelling through a crystal is usually split up into two rays polarised in perpendicular planes. If on emergence these rays travel along the same path they are in a fit state to combine, for they have the same period and travel through air with the same velocity. If they have been produced by an incident ray of ordinary light their relative intensity and phase difference may change every moment as the direction of vibration changes in the incident light, but if the incident light is itself polarised the relative intensity and phase difference of the rays remain constant at any one point within the crystal. Consequently on emergence the two polarised rays have a definite difference of phase and of intensity, and so will combine to form a simple polarised ray.

Interference of Light Rays.

It is obvious that two or more rays can in some circumstances travel along the same path without interfering with one another in any way. *A* can see *B*'s eye although at the same time *B* may be receiving rays from *A*'s eye. Also it is possible to see a reflection from a glass surface superimposed on an object seen through the glass; the particles carrying the light vibrations can be affected simultaneously by two sets of waves without detriment to either set. In the case of polarised light, however, if the period of the vibrations and the velocity of the rays are both constant, two coincident rays polarised in the same plane may show interference with one another. Each particle along the path will receive a pull up or down along the same direction due to each set of vibrations. Consequently one set of pulls may interfere with the other set, and the result may be a completely new train of waves.

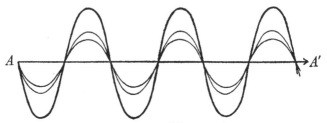

Fig. 87. Phase difference $=\lambda$.

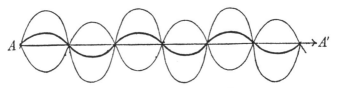

Fig. 88. Phase difference $=\dfrac{\lambda}{2}$.

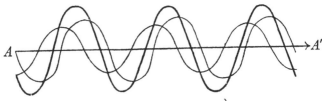

Fig. 89. Phase difference $=\dfrac{\lambda}{4}$.

Three cases may be considered, illustrated in figs. 87–9.

(a) Suppose that the two sets of vibrations are in the same phase; each particle in the path will receive two pulls in the same direction and will therefore be displaced to a greater degree than by either of the trains of waves occurring separately. In fact, its total displacement will be the sum of the displacements that it would undergo from the separate trains, and the combined effect will be a vibration of increased intensity. Fig. 87 gives a graphical representation of two rays reinforcing each other in this way. The two separate trains of waves are represented by thin lines, the resultant wave by a thicker line. Exactly the same result is obtained if two sets of waves of the same period travel along the same path with a phase difference equal to any number of complete wave-lengths.

(b) Suppose now that one train of waves is retarded so that it starts from A half a wave-length behind the other train (fig. 88). Under the influence of one train particle A is about to be pulled vertically upwards from its position of rest, while the other train is about to pull it vertically downwards. The resultant motion of A will be greatly diminished, and the amplitude of the vibrations will be reduced to the difference in amplitude between the two sets of waves. If these are equal in amplitude the forces acting on each particle at any moment of time will be equal and opposite, one vibration will entirely wipe out the effect of the other, and the two rays of light will together produce darkness. This will occur whenever there is a phase difference of an odd number of half wave-lengths between two otherwise similar trains of waves.

(c) Intermediate results occur when the phase difference between the two rays is some fraction of a wavelength other than $\frac{\lambda}{2}$. The resultant displacement at any point along the path can be found by taking the algebraic sum of the displacements due to the two separate waves in the same way as was done in the preceding examples. A regular train of waves of equal period but of different phase and different amplitude will be produced, as shown in fig. 89.

It must be clearly understood that interference is never produced between two rays of ordinary light or between rays polarised in different planes. The necessary conditions for interference are that the rays should be of equal wavelength (or period), polarised in the same plane and travelling at the same rate along the same path. Rays from different sources cannot be made so accurately of opposite phase that they produce complete darkness by interference, but this may be produced in practice by so arranging that two rays are derived from one source, one ray being made to lose phase in relation to the other to the extent of an odd number of half wave-lengths. A polarised ray passing through a crystal may be divided into two rays travelling with different velocities which may interfere on emergence from the crystal; hence the phenomenon of interference is of great importance in dealing with the optical behaviour of minerals.

One more point must be noted. Since interference only occurs between rays travelling with the same velocity and having the same period, rays of different colours can never interfere. But if white light is used it is possible that the conditions may be such that interference is produced for rays of a certain period, say the red rays, while for rays of another period, say the green rays, the conditions are such as produce reinforcement. In such circumstances the white light will be deprived of its red rays and perhaps to some extent of other colours also by partial interference, and it will take on the complementary tint, blue-green.

Such tints are known as *interference tints*; they afford a very useful guide to the study of minerals, and they will be treated more fully at a later stage (see p. 157).

REFLECTION, ABSORPTION AND REFRACTION

For all purposes relevant to the study of minerals light may be said to travel in straight lines. Just as a train of water waves moves along in a straight line unless it impinges on any obstacle, so the waves of light pursue a straight path unless interfered with by some obstruction. If, however, the rays encounter a different medium, either of two things may happen: the light may be turned back or *reflected* from the second medium, or it may partly enter into it. In the second case it is said to be *refracted*, and unless it is immediately absorbed, as it is by the particles of an opaque substance, the refracted light passes on through the second medium in a straight line which usually makes an angle with its original path. Only lack of homogeneity in the medium can make the path of a ray other than a straight line.

Reflection.

It is only by virtue of the light reflected from a non-luminous body that we are able to see it. Usually the surface of a body is rough, and although the roughness may be on an extremely small scale, it is generally not negligible in comparison with the very small wave-length of light, so that the reflected rays are thrown back from the object in all directions; this is known as *diffuse reflection*, and makes it possible for an object to be seen from many directions although it is only illuminated by light from one source. If a junction surface between two media is truly plane, a ray striking it is reflected back along one direction only, and the *Laws of Reflection* governing such reflection are the following:

(1) The incident ray, the reflected ray and the normal to the surface at the point of contact are all in one plane; this is known as the *plane of incidence*.

(2) The incident ray and the reflected ray make equal angles with the normal at the point of contact, the rays being on opposite sides of the normal. In other words, the angle of incidence is equal to the angle of reflection.

Diffuse reflection may be regarded as regular reflection taking place from very many small elements of surface which are not co-planar, so that the angle of incidence, and hence also the angle of reflection, varies from point to point. The reflected beam is always found to be less intense than the incident beam: we do in truth "see through a glass darkly"; some of the energy of vibration is not reflected but actually enters the reflecting surface. Moreover, it is found that the reflected light is to some extent polarised in the plane of incidence. The vibrations parallel to the reflecting surface have been reflected to a considerable extent, but vibrations in the plane of incidence to a much smaller extent. The amount of polarisation varies with the angle of incidence, and for any two media an angle can be found which gives the maximum amount of polarisation. In the case of reflection from an air/glass surface this angle is about $55\frac{1}{2}°$.

Absorption.

It has just been stated that the energy of a beam meeting a junction surface between two media is never entirely reflected back into the first medium; some of the energy is spent in producing light vibrations in the second medium through which they may continue to travel or in which they may be gradually absorbed. Often light of different wave-lengths is absorbed to different extents, so that any light which succeeds in passing right through the second medium is of a different colour from the incident light. This is the cause of ordinary coloration in transparent bodies. In some cases, as for example in some coloured glass bottles, the colour seen by reflected light is different from that seen by light passing through the body. Light of certain wave-lengths is almost wholly reflected from the surface of the bottle, while vibrations of other wave-lengths pass into the glass; part of this refracted light is absorbed and the residue gives the characteristic colour to the emergent light, which consequently is quite different from that of the reflected light.

The absorption of crystals will be studied in greater detail under the heading of *pleochroism* (see p. 154).

Refraction.

When a ray passes from one medium to another it usually undergoes a change in direction, and it is said to be refracted. The *Laws of Refraction* governing this change of path are as follows:

(1) The incident ray, the refracted ray and the normal to the surface at the point of contact are all in one plane, the *plane of incidence*.

(2) The incident ray and the refracted ray make with the normal to the surface angles whose sines are in a constant ratio for any two specified media.

If i is the angle between the incident ray and the normal, and r the angle between the refracted ray and the normal, then sin $i = \mu \sin r$, where μ is a constant for the two media whatever is the value of i. If the ray is passing from air to any other medium, the value of μ gives what is known as the *refractive index* of that medium.

Suppose that ab and $a'b'$ (fig. 90) are the limiting rays of

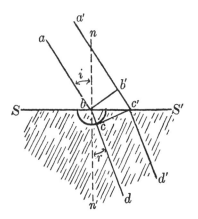

Fig. 90. Refraction of rays.

a beam of parallel rays travelling through air, and suppose that these rays strike a glass surface in b and c'.

The wave-front of the beam in air will be perpendicular to ab and $a'b'$; bb' represents its position when the ray ab strikes the glass surface. While ray $a'b'$ travels on through the air to strike the surface in c', ray ab travels through the glass in which it moves with decreased velocity, so that while $a'b'$ travels the distance $b'c'$, ab will have travelled a smaller distance bc, where

$$\frac{bc}{b'c'} = \frac{\text{velocity in glass}}{\text{velocity in air}}.$$

If therefore we draw round b a circle with radius bc, we know that the wave-front through c' must pass through some point on this circle, and also that it must be perpendicular to the path of the ray ab within the glass. Hence if we draw $c'c$ tangential to the circle, cc' will represent the wave-front and bc the path of the refracted ray ab inside the glass.

The refractive index of the glass is $\dfrac{\sin i}{\sin r} = \dfrac{\sin abn}{\sin cbn'}$. It is easy to see that this ratio of sines is also equal to the ratio $\dfrac{\text{velocity in air}}{\text{velocity in glass}}$, so that the ratio of the velocities is also a measure of the refractive index.

If the angle of incidence from air to glass is 0°, that is to say, if the incident ray is normal to the surface, $\sin i$ is zero, and hence $\sin r$ must be zero; in this case, therefore, there is no change of path on refraction.

Suppose the path of the ray to be reversed, and, starting in the glass, to travel out into the air; it is obvious that if cb is its path in the glass, ba will be its path in the air, and, unless they are both normal to the surface, the angle of refraction abn is greater than the angle of incidence cbn'. If the angle cbn' is gradually increased, at a certain inclination to the normal ba will become parallel to the glass/air surface. Rays in the glass still more inclined to the normal will not pass out into the air at all but will suffer reflection at the glass/air surface. This phenomenon, which is known as *total reflection*, can only take place when light is travelling from one medium into another of lower refractive index, and the limiting value of the angle of incidence which produces this effect is known as the *critical angle*.

Fig. 91 shows the refraction of rays at varying angles of incidence on a glass/air surface. No light from the region within the angle c_3bS' can penetrate into the air; it is all totally reflected back into the glass.

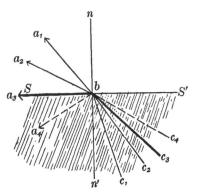

Fig. 91. Total reflection of rays.

In the case of the critical ray c_3ba_3, $\sin nba_3 = 1$; hence $\mu \sin c_3bn' = 1$; that is to say, the refractive index of the glass is the reciprocal of the sine of the critical angle. By measuring the value of the critical angle we can therefore deduce the refractive index of any medium denser than air.

REFRACTION IN CRYSTALS

When rays of light travel through space or through a medium whose particles are without definite arrangement, such as a gas or water or glass, there is nothing to give the vibrations a bias in this direction or in that; the rays travel through the medium in a perfectly normal way, and on meeting any surface they obey the laws of reflection and refraction. Media of this kind are described as *isotropic*;

since every direction in an isotropic medium is structurally the same as every other direction, the velocity of light travelling in the medium, and hence also the refractive index, is independent of the direction in which the light travels.

Only in the cubic system can crystals be said to possess the same structure in all directions, and cubic crystals can be grouped with amorphous substances as isotropic media*. The underlying reason is entirely different in the two cases, however; in cubic crystals the absence of any polarity imposed on the vibrations and of any alteration of velocity with alteration in direction is due to the perfect regularity of structure within the crystal, while amorphous substances, glass, water, air, etc., might be said to possess perfect irregularity of structure, any small groups of regularly arranged particles being counterbalanced by other small groups with a different orientation. In crystals belonging to all other systems there is a differentiation of atomic arrangement along different structural lines of the crystal, and this causes a variation of the velocity of rays according to their direction of vibration.

In general, a ray of ordinary light entering a crystal forms two distinct rays within the crystal which travel with different velocities and therefore are refracted to different extents and pursue different paths. Sometimes one ray follows the laws of simple refraction; sometimes neither does. Moreover, the two rays within the crystal are entirely polarised, each in a plane perpendicular to that of the other. This is the phenomenon known as *double refraction*

* This is only true so far as optical properties are concerned. Other properties, such as hardness, may vary in different directions even in a cubic crystal. See p. 93.

or *birefringence*. It may be regarded as a result of the regular arrangement of particles within a crystalline medium, the particles acting as a grid and restricting light vibrations to two directions of motion perpendicular to the direction in which the light is travelling. If the placement of particles is different along these two directions, that may be taken as sufficient explanation of the difference in velocity, and hence in refractive index, of the two polarised rays. If, on the other hand, there is no difference of orientation, we may expect there to be no difference of refraction between the two rays, and this is found sometimes to be the case. For example, if a ray travels along the vertical tetrad axis in a tetragonal crystal, the arrangement of particles is identical along any two mutually perpendicular lines in the plane of vibration; hence there is no *a priori* cause for double refraction, and the ray passes through the crystal without showing any signs of it. The same is true of rays travelling along the unique triad axis of rhombohedral crystals or the hexad axis of hexagonal crystals, but in no other case do crystals possess this regularity of structure about a structural line. Crystals belonging to these three systems, the tetragonal, hexagonal and rhombohedral, are therefore termed *uniaxial* crystals, and the direction of their unique axis of symmetry, along which there is only single refraction, is known as the *optic axis*. It should be remembered that the optic axis is a direction and not a line, and that light travelling in any part of the crystal or in any crystal medium is singly refracted as long as its direction is parallel to that of the optic axis of that crystal medium.

In crystals belonging to the three remaining systems it is not possible to select a direction perpendicular to which there is this regular arrangement of particles. These crystals do not possess a unique optic axis, but experiment shows that there are always two directions within the crystal in which light travels with single velocity. Crystals belonging to the rhombic, monoclinic and triclinic systems are therefore known as *biaxial* crystals. It is not immediately obvious from a study of space lattice models that biaxial crystals possess symmetry of structure round their optic axes, but if the structure and arrangement of molecules are taken into account it is possible to show that symmetry of a certain kind does exist.

Uniaxial Crystals.

Unless travelling along the optic axis, light entering a uniaxial crystal is split up into two beams polarised in perpendicular planes. One ray always vibrates perpendicularly to the optic axis, whatever its direction of propagation through the crystal, and its velocity and refractive index are also constant and independent of the direction of the ray. This is known as the *ordinary ray*; it obeys the laws of refraction, and in all ways behaves in a perfectly orthodox fashion. It is denoted by the symbol ω or o, and its velocity and refractive index are often referred to as V_0 and μ_0. The other ray is known as the *extraordinary ray*, and is often denoted by the symbol ϵ or e. Its plane of polarisation is perpendicular to that of the ordinary ray, so that its vibrations are always in a plane which also contains the direction of the optic axis. As the direction of vibration changes relatively to the optic axis, the orientation of crystal particles along the line of vibration will vary; consequently the velocity and refractive index of e varies according to the direction of the ray. When the ray is along the optic axis all vibrations must be perpendicular to the optic axis, and hence all travel with the velocity V_0. When the ray is only slightly inclined to the optic axis the velocity of the extraordinary ray is very nearly equal to that of the ordinary ray, but as the inclination of the ray increases, the difference in velocity increases until it reaches a maximum for rays travelling perpendicular to the optic axis.

In this case the vibrations of e are parallel to the axis, and the double refraction, as measured by the difference in refractive index between the ordinary and extraordinary rays, is at its maximum value. This value is taken as the strength of double refraction of the crystal in question.

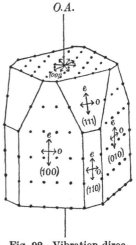

Fig. 92 attempts to show the relationship of double refraction to the crystal structure in a uniaxial crystal. It represents the top half of a tetragonal crystal showing the forms {100}, {001}, {110} and {111}. The dots show

Fig. 92. Vibration directions in uniaxial crystal.

the orientation of the particles lying in each plane, and the vibration directions of o and e are shown for a ray travelling through the crystal perpendicular to each of these faces. It will be seen that there is a simple relationship between these vibration directions and the pattern of dots lying in the plane of the face. Moreover, the vibration direction of o is always perpendicular to the direction of the optic axis, which here is vertical, while the vibration direction of e is parallel to the optic axis for the zone of faces (100), (110),

(010), and is inclined to it for vibrations in pyramidal faces of the type (111). In the face (001) the particles are arranged symmetrically about the optic axis, and light vibrates in this plane without any apparent polarisation and certainly without double refraction.

In the study of any mineral it is customary to cut thin sections of the crystal in different directions and to examine the behaviour of light passing perpendicularly through each section. For every section, except those perpendicular to an optic axis, there are two definite directions in its plane parallel to which the vibrations of the ordinary and extraordinary rays take place; these are usually known as the vibration traces for the slice. It is obvious from fig. 92 that in most cases the vibration traces are parallel to crystal edges. This is always so with pinacoidal and prismatic faces of uniaxial crystals, and even in the less regularly shaped pyramidal sections one vibration direction is usually parallel to a crystal edge*.

Of these two vibrations one always travels faster than the other; if the ordinary ray travels faster than the extraordinary ray, the crystal is said to be *optically positive*; if the ordinary ray is the slower ray, it is said to be *optically negative*. The reiteration of the letter o serves as a useful mnemonic in this connection:

if $V_o > V_e$, the crystal is optically positive;

while the e helps in a similar way for negative crystals:

if $V_e > V_o$, the crystal is negative.

It is interesting to note the shape of the wave-surfaces of the two rays. Supposing a point source of light within the

* The practical importance of this will be emphasised later under the subject of *extinction* (see p. 167).

crystal to emanate waves in all directions, the wave-surfaces of *o* and *e* will form complete curved envelopes. The ordinary ray travels with equal velocity in all directions; hence its wave-surface at any moment will be a sphere with the point source as its centre. The extraordinary ray travels with the same speed as the ordinary ray in the direction of the optic axis, since in this direction it is indistinguishable from the ordinary ray. In directions inclined to the axis

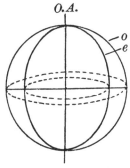

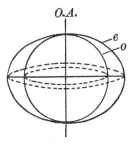

Fig. 93. Wave-surfaces for positive crystal.

Fig. 94. Wave-surfaces for negative crystal.

it travels at a varying velocity which reaches a maximum difference from that of the ordinary ray for directions perpendicular to the optic axis. The wave-surface for the extraordinary ray will therefore be a curved surface which has a circular section perpendicular to the optic axis and which touches the spherical ordinary wave-surface at two points lying on the line of the optic axis; it is the figure known as an *ellipsoid of revolution*. Figs. 93 and 94 show the two possible cases. In fig. 93 the velocity of *e* decreases as the ray becomes more inclined to the optic axis, until it reaches a minimum for rays travelling perpendicular to the axis;

this is the case of a positive crystal. The wave-surface of the extraordinary ray will be a regular egg-shaped ellipsoid wholly contained within the spherical wave-surface of the ordinary ray; the outline of the figure suggests the + sign of the crystal to which it belongs. In fig. 94 the velocity of the extraordinary ray is greater than that of the ordinary ray, and its wave-surface resembles that of a muffin-dish, wholly enclosing the spherical wave-surface. The elongated outline of the whole diagram again gives a mnemonic for the − sign of the crystal. It must be pointed out that the figures very greatly exaggerate the separation of the wave-surfaces. Even in calcite, a mineral with very high double refraction, the ratio $V_o : V_e = 1 : 1\cdot116$, while in quartz, which is optically positive, $V_o : V_e = 1 : 0\cdot994$.

In both positive and negative crystals rays travelling along or perpendicular to the optic axis are perpendicular to their wave-fronts, but for any other direction the extraordinary ray is not normal to its wave-surface.

Biaxial Crystals.

Crystals belonging to the rhombic, monoclinic and triclinic systems have no unique structural axis about which there is a symmetrical arrangement of particles. Correspondingly, they have no unique optical axis which is related to their structure and along which there is only single refraction; two directions can always be found, however, which in respect of giving single refraction are equivalent to the single axis of a uniaxial crystal. Crystals belonging to these three systems are therefore known as *biaxial crystals*.

When light enters a biaxial crystal it is usually divided into two rays polarised in perpendicular planes and possessing different velocities and refractive indices. As the direction of the ray varies its velocity varies also; there is no ray which can be termed "ordinary" in the sense that its velocity is constant and that it obeys the laws of refraction. Both rays are extraordinary, and in general they are refracted to different extents on entering the crystal. The surfaces of the two waves* which would spread out from a point source within the crystal take the form of a double surface somewhat resembling two slightly irregular ellipsoids which intersect in four points. The lines joining opposite points are co-planar; they pass through the central point of the ellipsoids, and each is therefore a direction along which rays belonging to both sets of waves travel outwards from the centre with equal velocities; that is to say, these lines are the two optic axes of the crystal.

It is difficult in a two-dimensional diagram to show the exact shape of these wave-surfaces; it is more satisfactory to consider instead their outlines as shown in three principal planes at right angles, the position of the planes being decided by the experimental determination of the refractive index for light travelling in various planes. As in the case of uniaxial crystals, the refractive index varies with the direction of vibration rather than with the direction of propagation. Two rays travelling along the same path in the crystal may have different refractive indices because they are polarised in perpendicular planes, but two rays travelling along entirely different paths will have the same velocity and refractive index as long as their vibration directions are the same. If, by methods to be described later, the refractive index of a crystal is measured for rays polarised in

* These are really ray-surfaces rather than wave-surfaces, since they are defined at any moment by the points reached at that moment by rays spreading out in all directions from the source.

different planes, it is found that there are in every biaxial crystal two mutually perpendicular vibration directions which yield maximum and minimum values for the refractive index. These values are denoted by γ and α respectively; the refractive index for vibrations in the third perpendicular direction is denoted by β, and is intermediate between the other two, but is seldom the arithmetic mean of α and γ. In describing the optical properties of a biaxial crystal the

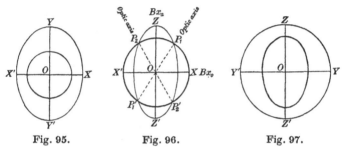

Fig. 95.   Fig. 96.   Fig. 97.

Figs. 95–97. Principal section planes of wave-surfaces in
biaxial crystal.

values of α, β and γ are always quoted, and from them the refractive index for vibrations in any other direction can be calculated.

Three perpendicular planes can therefore be chosen, each containing two of the vibration directions for which the refractive indices are α, β and γ; these are known as the *principal section planes* of the crystal.

Figs. 95–97 represent sections of the wave-surfaces in each of these three principal section planes. OX is the vibration direction for maximum velocity and minimum refractive index α, OY for mean values of each, and OZ for minimum velocity and maximum refractive index γ.

DC   9

In the plane OX, OY (fig. 95), rays vibrating perpendicular to the paper and therefore parallel to OZ will spread out from O with equal velocity in all directions, and this velocity will be the minimum for the crystal. Hence the section of one wave-surface will be a circle whose radius is proportional to the minimum velocity. Another ray will travel from O towards X vibrating parallel to OY, and hence moving with mean velocity. As the direction of this ray moves round towards OY, its vibration direction will remain in the plane of the paper but will change from OY to OX, and at the same time its velocity will change from the mean to the maximum value. Hence the section of the second wave-front will be an ellipse with semi-axes proportional to the mean and maximum velocities, and this section will entirely enclose the circular wave-front, which travels with the minimum velocity for the crystal.

Similarly, in the plane OY, OZ, the section of one wave-surface is circular, with a radius proportional to the maximum velocity, and that of the other is elliptical, as shown in fig. 97, and these surfaces do not intersect.

In the plane OX, OZ (fig. 96), the two wave-surfaces intersect at the extremities of two lines P_1OP_1' and P_2OP_2'; these are the two co-planar lines referred to before along which two rays emanating from a point source at O travel with equal velocity, although vibrating in different planes. P_1OP_1' and P_2OP_2' are therefore optic axes of the crystal. The wave-surfaces do not intersect in any other points, so that in crystals of this type there are two and only two optic axes. It must be noted that they lie in the plane of the maximum and minimum velocity vibration directions OX and OZ, which in consequence is known as the *optic axial plane*, and that they are equally inclined to OX and

OZ. One of these directions bisects the acute angle between the optic axes, and is known as the *acute bisectrix* or Bx_a, while the other, which bisects the obtuse angle between them, is known as the *obtuse bisectrix* or Bx_o. The acute angle between the optic axes is denoted by $2V$, and the value of this is quoted in the description of the optical properties of a crystal*. The single refractive index for all rays travelling along the optic axis is the mean refractive index β.

In a complete analysis of the optical properties of a biaxial crystal it is necessary to distinguish between OP_1 and OP_2, the axes of single *ray* velocity, and two other directions nearly coincident with them which represent directions of single *wave* velocity. The velocity of a wave must be measured along a line normal to itself; OP_1 is normal to the circular wave-section but is inclined to the elliptical wave-section; the perpendicular to both sections is OM_1, where OM_1 is perpendicular to the tangent common to both circle and ellipse (fig. 98). A wave-surface, travelling through

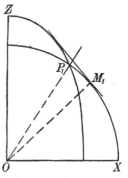

Fig. 98. Primary and secondary optic axes.

air parallel to this tangent, will, on entering the crystal, be divided into two wave-surfaces, formed by rays vibrating in perpendicular planes and represented in the section plane OX, OZ by the circle and the ellipse. The

* The angle $2V$ between two rays travelling along the optic axes in a crystal will by refraction into air at a surface normal to the Bx_a be increased to a greater angle, and this apparent divergence of the optic axes in air is denoted by $2E$. It is obvious that $\sin E = \beta \sin V$.

direction of propagation of a wave-surface at any point is perpendicular to the tangent to the surface at that point; hence the wave-surfaces inside the crystal will proceed without change of direction along the line OM_1, the normal to the common tangent, and this direction can therefore be termed the direction of single refractive index and single velocity for wave-surfaces. OM_1 and its fellow axis OM_2 in the neighbourhood of OP_2 are known as the *primary optic axes*, while OP_1 and OP_2, the axes of single ray velocity, are known as the *secondary optic axes*. In practice the primary and secondary axes are so nearly coincident that it is unnecessary for most purposes to differentiate between them; in aragonite, the rhombic form of $CaCO_3$, the angle between them is only 2°, and this is an unusually high value.

In some respects the acute bisectrix may be regarded as the counterpart of the single optic axis of uniaxial crystals; if the two axes of a biaxial crystal were made to approach each other more and more nearly, they would finally coincide in the acute bisectrix, which then would become the optic axis of the resulting uniaxial crystal. Bearing this in mind, the following convention as to optical sign in biaxial crystals falls into line with that for uniaxial crystals. In a positive uniaxial crystal the optic axis is the direction of vibration of rays having the maximum refractive index; in positive biaxial crystals the vibration direction for rays having the maximum refractive index γ is the acute bisectrix. Another way of stating this convention is that in a positive biaxial crystal $\gamma - \beta > \beta - \alpha$; that is to say, the value of the mean refractive index β is nearer to that of the minimum index α than to that of the maximum index. The identity of the two statements is not perhaps immediately obvious, but it will become simpler after a consideration of the *optical ellipsoid*, a figure constructed to represent in a comprehensive and quantitative way the optical properties of a crystal.

The Optical Ellipsoid.

It has been seen that the morphological characteristics of a crystal, its symmetry and its interfacial angles, are summarised comprehensively by stating the nature of the crystallographic axes, their relative lengths and their inclinations. In somewhat the same way the optical properties of a crystal can be summarised by a statement of the geometrical constants of a solid figure known as the *optical ellipsoid.* An ellipsoid is a curved figure of three dimensions built up on three perpendicular axes. If two of these axes are equal, the figure becomes what is known as an ellipsoid of revolution, for its surface could be described by the rotation of an ellipse about one of its axes, which then becomes the unequal axis of the ellipsoid. If all three axes are equal, the ellipsoid becomes a sphere. In general, a section through the centre of an ellipsoid is an ellipse with a major and a minor axis at right angles to each other. There are, however, two sections which are circular; these are equally inclined to the greatest and to the least axis of the ellipsoid, and they contain the third axis. In an ellipsoid of revolution there is one circular section only, the principal section which contains the two equal axes.

There are two optical ellipsoids in common use. The first one was suggested by Fresnel, who took as each axis of the ellipsoid a length proportional to the velocity of light vibrating parallel to that axis. The length of any radius of the ellipsoid represents the velocity of a ray vibrating

parallel to that radius and travelling in any direction
perpendicular to it. This ellipsoid is sometimes known as
the *vibration-velocity ellipsoid.*

The second ellipsoid is that due to Fletcher and described
by him in 1891. His axes are the reciprocals of those of
Fresnel's ellipsoid, being proportional to the refractive
indices α, β and γ of light vibrating parallel to the axes.
This ellipsoid is known as the *optical indicatrix.*

In a given crystal the positions of the axes are the same
for both ellipsoids, and the three principal sections of
the ellipsoids, each containing two of the axes, coincide
with the principal sections of the wave-surfaces already
considered (p. 129). The major axis of one ellipsoid
coincides with the minor axis of the other, however, and
vice versa.

Summarising the properties of the Fresnel ellipsoid,
suppose (fig. 99) that OP is any direction in the crystal,

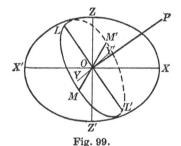

Fig. 99.

and that LL' and MM' are the principal axes of the
section normal to OP. Then OL and OM represent the
vibration directions of the two plane polarised rays that
can travel through the crystal along the line OP, and their
lengths are proportional to the velocities of those rays. If

now OP is perpendicular to a circular section of the ellipsoid, $OM = OL$, and all rays will travel along OP with constant velocity regardless of their plane of polarisation. In
this case OP is the position of the secondary optic axis, the
axis of single ray velocity. We may therefore define the
optic axis as the normal to the circular section of the optical
ellipsoid. It must be remembered that OP is not generally
the direction of single wave velocity also, since the wave-
front within a doubly refracting crystal is not generally
perpendicular to the rays which form it. When the rays are
travelling along any axis of the ellipsoid they are normal to
the wave-front, but otherwise they make with it an angle
which differs slightly from 90°.

In the case of a uniaxial crystal, two of the ellipsoidal
axes become equal, and the third axis becomes the single
optic axis of the crystal. In a positive crystal the velocity
of the ordinary ray, vibrating perpendicularly to the optic
axis, is greater than that of the extraordinary ray. The
vibration-velocity ellipsoid of a positive uniaxial crystal
therefore has two equal axes which are greater than the
third axis, and the ellipsoid is of the muffin-dish form
(fig. 100). In a negative uniaxial crystal the two equal axes
are shorter than the third axis, and the vibration-velocity
ellipsoid resembles an egg with exactly similar ends
(fig. 101). In each case OZ represents the position of the
optic axis.

In the case of a biaxial crystal suppose that OX and OZ
(fig. 102) represent the maximum and minimum velocity
vibration directions. There must be a radius of the ellipsoid in the plane XOZ intermediate between OX and OZ
which is equal in length to OY: call this OR_1. The section
through OY and OR_1 will be a circular section, and OP_1,

the normal to this section, will therefore be an optic axis.
There will be a second radius OR_2 between OZ and OX' which

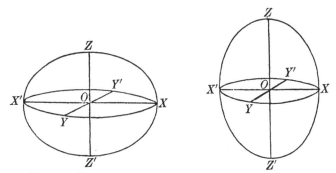

Fig. 100. Positive crystal.   Fig. 101. Negative crystal.
Vibration-velocity ellipsoids for uniaxial crystals.

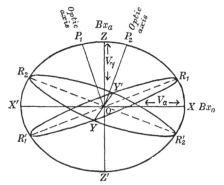

Fig. 102. Vibration-velocity ellipsoid for positive
biaxial crystal.

is equal to OY, and consequently there will be a second
circular section and a second optic axis OP_2, and the two
circular sections and the two optic axes will be symmetrical
about OZ. In the case shown OZ, the vibration direction for

minimum velocity, is the acute bisectrix of the angle $2V$ between the optic axes; OZ corresponds to the direction of maximum refractive index γ, so that the direction of γ is the Bx_a and by convention the crystal is regarded as optically positive.

Side by side with Fresnel's vibration-velocity ellipsoid we may put Fletcher's indicatrix for the same crystal (fig. 103). OX now becomes the smallest axis and OZ the greatest, since they represent the vibration directions and the relative values of the minimum and maximum refractive indices α and γ respectively. The ellipsoid will yield two circular sections, the normals to which will lie in the plane of α and γ. If β is nearer in value to α than to γ, which was the first statement of the convention for an optically

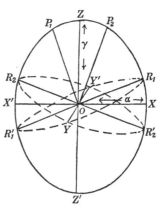

Fig. 103. Fletcher's indicatrix for a positive biaxial crystal.

positive crystal, the plane of the circular section will be nearer to OX than to OZ; hence the normal OP_1 will be nearer to OZ than to OX. This means that OZ, or γ, will be the acute bisectrix, which was by the second statement of the convention (p. 132) the condition for an optically positive crystal.

In Fletcher's indicatrix OP_1 and OP_2 are the directions which yield a single refractive index only. Hence light travelling along OP_1 or OP_2 will undergo the same amount of refraction in passing out into the air, or from the air into the crystal—that is to say, the wave-front of a beam of light travelling from air into the crystal will undergo refraction in the formation of both polarised beams within the crystal, so that these beams possess parallel wave-fronts. Hence in the indicatrix OP_1 and OP_2 are the axes of single wave velocity, or the primary optic axes, and so, theoretically at any rate, not quite coincident with the secondary optic axes of Fresnel's ellipsoid.

For any elliptical section of Fletcher's indicatrix, the normal to the section is the direction of two waves (not usually of two rays) which are polarised so that their vibrations are in the directions of the two axes of the elliptical section, and whose refractive indices are represented by the lengths of these axes. The difference in length of these axes gives a measure of the double refraction for a plate of the crystal cut parallel to this section; the maximum value of the double refraction is given by the section OX, OZ, and is represented by the value of $\gamma-\alpha$.

It is only along the three axes of the ellipsoid that a ray entering the crystal normally to the surface of the ellipsoid is not bent out of its course but passes through the centre of the ellipsoid. In that case the section normal to the path of the ray is a principal section; the lengths of the semi-axes give a measure of the refractive indices of the two polarised rays, and their positions give the directions of vibrations of the rays. In the general case, however, the ray is not normal to the surface of the ellipsoid, and the two polarised rays within the crystal, having different refractive indices, travel along different paths, so that the axes of the elliptical section no longer give exact data as to their properties. Thus, if OP (fig. 104) is any ray passing through the centre O of the ellipsoid, the refractive index of the ray is not given by the length of the axis OR of the elliptical section normal to OP, but by the length MN, where MN is normal to OP and to the surface of the ellipsoid. Two such normals will be possible, the other perpendicular to the paper, and these two lines represent the refractive indices of the two polarised rays that can travel along OP and their directions of vibration. It is obvious that when OP is an axis of the ellipsoid, OR, the normal to OP, will also be perpendicular to the tangent, so that the axes of the section will then give the data as to refractive index and plane of polarisation.

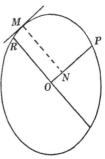

Fig. 104.

Orientation of Optical Ellipsoid with regard to Crystallographic Axes.

The optical ellipsoid has a definite orientation with regard to the structural axes of a crystal, although in the less symmetrical systems there are certain degrees of freedom in their relationship.

In cubic crystals the ellipsoid becomes a sphere; its axes are all equal, and there is no difference in velocity or refractive index for rays polarised in different planes.

In uniaxial crystals, which belong to the tetragonal, hexagonal or rhombohedral systems, the ellipsoid is one of revolution. The unique axis of the ellipsoid is coincident with the unique axis of symmetry, which is also the vertical crystallographic axis C, so that this is also the optic axis. The two ellipsoidal axes perpendicular to this are equal, so that light vibrating in their plane, and therefore travelling along the unique axis, suffers only single refraction. The maximum double refraction is given by the difference in length between the axes of the ellipsoid, and is shown by the two rays travelling perpendicular to the unique axis.

The orientation of the ellipsoid in biaxial crystals is less definite.

In rhombic crystals the axes of the ellipsoid coincide with the crystallographic axes, but there is no steady relationship between the crystallographic axial ratios and those of the ellipsoid. The optic axial plane is therefore one of the three principal planes of the crystal. Any one of the crystallographic axes may be the acute bisectrix, depending on the original orientation of the crystal (see p. 49), and the maximum double refraction is shown by light travelling perpendicular to the optic axial plane.

In monoclinic crystals one axis of the ellipsoid coincides with the crystallographic axis *B*, which is also the diad axis of symmetry, while the other two axes of the ellipsoid may lie anywhere in the symmetry plane containing the *A* and *C* crystallographic axes. The optic axial plane may therefore be this plane of symmetry, the axes and the bisectrices lying anywhere within it, in which case the diad axis *B* is the vibration direction of mean velocity and refractive index; or the optic axes may be in any plane through the diad axis perpendicular to the plane of symmetry, in which case the diad axis is coincident with one of the bisectrices, and is the vibration direction of maximum or minimum refractive index. The ellipsoid is, so to speak, pivoted by one of its axes along the diad axis of the crystal, but can rotate about that axis and take up any position according to the structure of the particular crystal.

In triclinic crystals all restriction of orientation of the ellipsoid has vanished; its axes may run in any direction through the crystal, and its position can only be determined by finding experimentally the vibration directions for rays possessing maximum and minimum refractive index.

Dispersion of Optic Axes and Bisectrices.

In all cases where an optic axis or bisectrix is not coincident with a crystallographic axis, its position may vary with the wave-length of the light used. This produces what is known as *dispersion of the optic axes and bisectrices*. No dispersion is possible in uniaxial crystals. In rhombic crystals there can be no dispersion of the bisectrices since they are coincident with the crystallographic axes, but the axial angle $2V$ may be slightly different for different wave-lengths, and there may be consequent dispersion of the axes within the optic axial plane. This is usually only small, but in a few cases the phenomenon is very pronounced. Brookite, a rhombic form of titanium oxide, is especially remarkable in that for red light the axial plane is parallel to the basal plane (001), the A axis being the acute bisectrix, and the axial angle is 58°, while for yellow sodium light the axial angle is reduced to 38°. As the wave-length decreases further the axial angle also decreases, until for a certain wave-length of greenish-yellow light the crystal becomes uniaxial, the A axis being the direction of the single optic axis. For still shorter wave-lengths the optic axes open out again, but this time in a plane parallel to (010), so that for green light (010) is the optic axial plane, the A axis is the acute bisectrix and the optic axial angle is $21\frac{1}{2}°$.

Monoclinic crystals show dispersion of the optic axes and

of one or both of the bisectrices according to the orientation of the ellipsoid, while triclinic crystals can show dispersion of all four lines. In all cases the colour dispersion is symmetrical about those axes of the ellipsoid which coincide with crystallographic axes and therefore cannot themselves show dispersion.

Similar changes in the positions of the axes or bisectrices may be produced by change of temperature.

CHAPTER VI

METHODS OF INVESTIGATION OF OPTICAL PROPERTIES

Although for some processes of investigation the mineral is used in its original shape, or cut into prisms, the most widely useful form for general optical examination is a thin slice with parallel sides about 0·03 mm. thick, mounted in Canada balsam on a microscope slide*. The slice is thin enough to render transparent many minerals which are opaque in massive form, and this facilitates the study of their optical properties. When such a section is placed on the stage of a microscope, light enters it normally from below and emerges normally from its upper surface, and any change in the nature of the emergent rays can be attributed to the behaviour of the crystal with regard to light impinging upon it in a direction normal to the plane of the section. For a complete optical examination several sections must be cut in various directions; in the following description of the phenomena observed reference will be made to crystals of various systems cut along specific planes.

PETROLOGICAL MICROSCOPE

For use with minerals a microscope must have certain fittings in addition to the fundamental lenses constituting the optical system. It must possess a mirror or other

---

* Directions for the preparation of thin slices of rocks or mineral may be found in the text-books of Dana or Miers, or in Johannsen's *Manual of Petrographic Methods.*

arrangement to send a sufficient supply of light up into the
tube of the instrument, and a lens or combination of lenses
below the stage, known as the *condenser*, which can be
swung in and out of action and which serves to focus the
light on to the centre of the slide. The stage must be
capable of rotation about the axis of the instrument and
must be graduated round the edge, the scale working
against a fixed pointer or vernier. Finally, an indispensable
feature of the petrological microscope is an arrangement
for polarising the light incident on the slide, together with
a means for the analysis of the light after it has passed
through the mineral section. Both these purposes are
served by the *nicol prism*, and every petrological microscope is fitted with two of these, the lower one, or *polariser*,
being fitted below the condenser, and the upper one, or
analyser, in some position in the tube of the instrument;
each can be put in or out of action at will.

NICOL PRISM

The nicol prism plays such an important part in the optical
examination of minerals that it must be described in some
detail. It is constructed from a crystal of calcite whose
length is about three times its breadth. Calcite in the form
known as Iceland spar occurs in blunt rhombohedra.
A perfectly transparent elongated rhomb is chosen (fig. 105)
of such dimensions that the shaded diagonal plane cutting
the crystal symmetrically through the blunt corners C and
A' is at about 90° to the end faces $ABCD$ and $A'B'C'D'$.
The crystal is cut through along this plane; the cut surfaces
are polished and re-cemented, leaving a very thin parallel-
sided layer of Canada balsam between them. The end faces
are ground down and polished so that they make an angle

of 68° with the vertical sides; the latter are coated with black, and the crystal is mounted in a metal cylinder, so that the only parts exposed are the rhombic faces *ABCD* and *A'B'C'D'*.

The unique triad axis of symmetry, which is also the direction of the optic axis, runs through the blunt corner *C* in a direction *OO'* which is equally inclined to the three

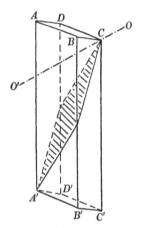

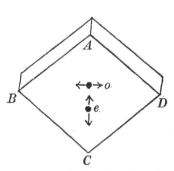

Fig. 105. Construction of nicol prism.

Fig. 106. Refraction through calcite rhomb.

crystal faces meeting in *C*, and which lies in the diagonal plane *AA'C'C*. Parallel to *OO'* there is single refraction, but parallel to the length of the crystal the birefringence is so strong that a black dot viewed through quite a small thickness of the crystal gives a double image (fig. 106). One image, *o*, which remains still as the crystal is rotated about the line of sight, is formed by the ordinary rays which vibrate perpendicularly to the optic axis; the second image, *e*, rotates with the rotation of the crystal, the two images always lying along the shorter diagonal of the face *ABCD*. Image *e* is formed by the extraordinary rays whose vibrations are in the same plane as the optic axis, that is to say, along the shorter diagonal of the face *ABCD*.

Suppose *AA'C'C* (fig. 107) is a vertical cross-section of the crystal after it has been cut diagonally and re-cemented.

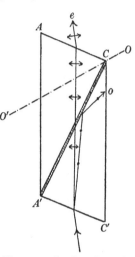

CA' represents the trace of the Canada balsam layer, and *OO'* the optic axis which lies in the plane *AA' C'C*. Light entering the crystal at the lower surface *A'C'* will be doubly refracted; of the two rays formed *o* has the refractive index 1·658, which is considerably greater than the index 1·540 of Canada balsam, while *e* for this direction has an index 1·536 nearly equal to that of the balsam. The dimensions of the prism are such that *o* strikes the Canada balsam at an angle of incidence greater than the critical angle; consequently *o* is totally reflected towards the

Fig. 107. Section through nicol prism.

side of the prism and is absorbed at the blackened surface. The extraordinary ray, vibrating in the plane of the optic axis, which in fig. 107 is the plane of the paper, passes through the slightly more highly refracting balsam and finally emerges from the top surface *ABCD*. Hence all the light issuing from the prism is entirely plane polarised, and its vibrations are parallel to the shorter diagonal of the face *ABCD*.

If the nicol prism which forms the polariser of a petrological microscope is swung into place below the stage, the light that reaches the stage is plane polarised in a direction that can easily be found by inspecting the top face of the prism. If the polarising and analysing nicols are set so that the shorter diagonals of their exposed faces are parallel, the rays coming through the polariser will pass without change through the analyser. If, however, the analyser is slowly rotated about the line of sight, the intensity of the light diminishes until, when the shorter diagonal of the analyser is at right angles to that of the polariser, all the light passing through the latter is vibrating in the direction of the ordinary ray in the upper nicol, and so is entirely cut off. Two nicols when in this relatively perpendicular position are said to be *crossed*, and together they transmit no light at all. In positions intermediate between the crossed and the parallel positions light from the polariser is doubly refracted on entering the analyser, and only that portion is transmitted which forms the extraordinary ray in the second nicol. The light is therefore noticeably diminished in intensity but not entirely cut off.

MICROSCOPIC EXAMINATION OF A MINERAL

The examination of a mineral by means of the microscope may be divided into four sections:

A. *Examination in ordinary light*. For this the polariser and analyser are slid out of the line of vision; the mirror is adjusted to send a beam of parallel light up through the tube of the instrument, or, if a convergent beam is required, the condenser is swung into place. The microscope is used in this way for the study of the shape of crystal sections, their colour and other characteristics such as cleavage traces, state of preservation, inclusions, and, in the case of rock slices, the relationship of one mineral to another. Finally, comparative or absolute estimates of refractive index can also be obtained.

B. *Examination in plane polarised light.* The polariser is slid into place, and by rotating the stage the appearance of the mineral can be studied in light vibrating in various directions relative to its crystallographic structure. The chief phenomenon which is revealed by this arrangement is that of pleochroism.

C. *Examination between crossed nicols.* The polariser and the analyser are both slid into the line of vision, having previously been adjusted to the " crossed " position. Information can then be obtained with regard to double refraction, the directions of vibration in a mineral section, and the optical sign of the mineral.

D. *Examination in convergent polarised light.* The converging lens is placed between the polariser and the section, a high-power object-glass is used, and a second lens is inserted to enable the rays to be focussed by the eye-piece. Alternatively, the eye-piece is removed, and the so-called interference figure which is produced by the convergent rays passing through the section is viewed directly by the eye. Interference figures are used to distinguish between sections of uniaxial and biaxial crystals and to determine the optical sign of a crystal.

A. *Examination in Ordinary Light.*

General appearance. A certain amount of information can be obtained from the general appearance of crystal sections under the microscope. The outline is sometimes sufficient to suggest the system to which a crystal belongs, especially in the case of cubic and hexagonal crystals. A few minerals, notably metalliferous ores, remain opaque even in very thin slices, others show characteristic colour, while the majority appear colourless. By the relative perfection of crystal outline and the impression received of one mineral embedded within another, it is possible to determine the order of crystallisation of the various constituents in a rock slice, while a lack of clearness often denotes that chemical decomposition has taken place to some extent. Cleavage is revealed as a series of more or less regular lines parallel usually to one side of the crystal section, and its quality and directions can be noted.

Refractive index. The fundamental optical property of a substance is its refractive index. In all crystals except those of the cubic system the refractive index varies with the direction of the light vibrations, and, moreover, in almost all media it varies with the wave-length of the light. For accurate work, therefore, it is necessary to specify the particular wave-length for which the refractive index has been measured; where this is not stated it may be assumed that the value quoted is that for the yellow light of the sodium flame. Since the wave-length of yellow light is roughly the average of those of all visible rays, the corresponding refractive index is a medium value and agrees fairly closely with the value obtained by less accurate methods using white light. For the petrologist's purpose values of refractive index correct to two places of decimals are usually sufficient. For uniaxial minerals the values for the ordinary and extraordinary rays must be determined, and for biaxial minerals the three values α, β and γ. The difference between the two extreme indices, which denotes the strength of double refraction or birefringence, varies considerably in different minerals, but it is not often in excess of 0·1. The maximum refractive index itself varies between 1·4 and 1·8, and it is possible to describe a mineral accordingly as of low refractive index (e.g. fluor spar, 1·43), of medium refractive index (e.g. quartz, 1·54–1·55, and the felspars, which vary between 1·51 and 1·58 according to their composition), or of high refractive index (e.g. biotite, 1·57–1·63, and hornblende, 1·63–1·65). The refractive index of olivine has the very high value of 1·69, while garnet is quite outstanding with a refractive index which in some species may reach 1·80. There is no direct relationship between strength of refraction and strength of birefringence; quartz has a low birefringence of 0·009; white mica varies in different sections from weak birefringence of 0·01 to three times that amount; dark mica, biotite, is considered to be strongly birefringent with a maximum difference of 0·06 between its indices, while calcite shows the remarkably high value of 0·172.

The method employed for the measurement of refractive index depends upon the accuracy that is required. The index of a mineral relative to that of an adjacent mineral or of the Canada balsam ($\mu = 1\cdot54$), is revealed in a microscope slide by the *relief* exhibited by the mineral. The junction between two media of widely differing refractivity shows up in strong relief, while that between two media of very nearly the same refractivity, such as quartz and Canada balsam, is marked only by a fine, inconspicuous line. This effect is due to the phenomenon of total reflection which takes place at the junction of the two media; this has the result also of throwing into strong relief any cracks or surface irregularities of a highly refracting mineral such as garnet.

F. Becke made use of this effect to determine which of two adjoining minerals in a rock slice possessed the higher refractive index. Suppose that A and B (fig. 108) are two adjacent media in a rock slice, A being of the higher refractive index. Usually their junction will be oblique to the surface of the slice, and the two possible cases are shown in fig. 108. Parallel rays falling normally on the lower surface of the section will be deflected as indicated, and accordingly the distribution of light as seen through the microscope will

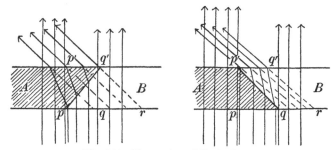

Fig. 108. Formation of Becke line.

not be even over the field. When the microscope is focussed on the lower surface the rays from the area pq are deflected and will appear to come from the area between q and r. From this area also will come the other rays which pass normally through B, so that the junction between A and B will show a bright line on the side of B and a dark line on the side of the more highly refracting substance A. If now the focus is raised to the upper surface the deflected rays appear to come from an area round about p', while little, if any, light comes from the area about q'*. Hence there will be a

* The effect will be the same if the difference in refractive index is sufficient to cause total reflection of rays from the A to B surface.

bright line along the junction on the side of A. That is to say, as the focus is moved *up*, the bright line, or Becke line as it is called, moves *up* towards the medium of higher refractive index. To see the line most easily a stop should be placed below the polariser so that the junction is lighted by a narrow pencil of parallel rays.

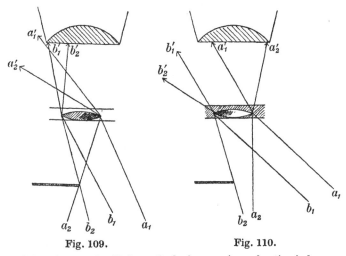

Fig. 109.   Fig. 110.

Schroeder van der Kolk method of measuring refractive index.

The method has been further developed by Schroeder van der Kolk for measuring the absolute value of refractive index. A small grain of the mineral is placed on a slide and surrounded with a liquid of known refractive index, oblique illumination being used. The grain is usually roughly lenticular in shape, and figs. 109 and 110 show the type of refraction which takes place when it is surrounded by liquids of lower and higher refractive index respectively. In both cases with oblique illumination light reaches the object lens from only one half of the grain, while the other half looks dark. The microscope reverses the image, so that if the grain has a higher refractive index than the surrounding liquid, as in fig. 109, the shadow appears to be on the opposite side to that of the source of light. The experiment is repeated with various liquids of graded refractive index, until the absence of any shadow indicates that the index of the liquid is equal to that of the grain. Yellow sodium light should be used for accurate measurement; if white light is used the greater difference between the refractive indices for red and for blue rays in the case of liquids produces colour effects, the bright part of the grain looking red when its index is greater than that of the liquid, and blue when the reverse is the case. The colours only appear when the difference in refractive index is very small, or when it is actually zero for yellow light but appreciable for red or blue rays.

Various liquids are used for the purpose, covering a wide range of refractive indices. For medium and low indices various oils can be used, such as pure castor oil (1·48), clove oil (1·54) or cinnamon oil (1·60), and any one of these can be mixed with monobromnaphthalene (1·66) or methylene iodide (1·74) to give a very finely graded series of refractive powers. The grain is placed on a slide with a cover-slip over it, a drop of the appropriate liquid is deposited against the edge of the slip and is drawn beneath it by surface tension, completely surrounding the grain.

Other methods can be used if larger specimens of the mineral are available. For practical purposes measurements are most easily made by means of some form of refractometer, the working principle of which is the phenomenon of total reflection. For a description of this and of other methods the reader is referred to the text-books of Dana, Miers, Tutton or Johannsen.

In the case of doubly refracting minerals determinations of refractive index have to be made with polarised light on carefully oriented sections or prisms cut from the crystal. In this way values may be obtained for the refractivity of the mineral with regard to light vibrations in certain specific directions, which can be chosen with special reference to the crystallographic structure, and from the maximum and minimum values obtained the optical ellipsoid for the crystal can be constructed.

B. *Examination in Plane Polarised Light.*

Pleochroism. By sliding the polarising nicol prism into the line of vision, the vibrations of the light falling on to the stage of the microscope are restricted to one direction; this may produce differential effects in sections of a mineral which influences rays of light differently according to their direction of vibration. The most marked effect to be noted is due to the phenomenon of *pleochroism*, or, in other words, of differential absorption for rays of different wave-lengths, the absorption varying according to the vibration direction of the light.

It has already been noted that the colour of a transparent body by transmitted light is due to the greater absorption of some wave-lengths than of others. Crystals which possess different values of refractive index according to the direction of vibration of the rays may be expected to show different amounts or types of absorption according to the direction of vibration, and the colour of light transmitted through a section of a birefringent mineral may vary according to the direction of vibration of the transmitted light. Suppose therefore that the section is viewed by plane polarised light; as the orientation of the section is changed relatively to the vibration direction of the polariser, the colour of the section may also change. More generally, a birefringent crystal may appear by transmitted light to have different colours when viewed in different directions, and it is this property that is known as pleochroism.

Isotropic minerals have equal absorptive powers in all directions. Uniaxial crystals show two extreme tints, one given by the ordinary rays vibrating perpendicular to the optic axis, and the other by the extraordinary rays vibrating parallel to the optic axis; extraordinary rays vibrating in any other direction give a colour intermediate between the two extreme tints. Hence a section of a uniaxial crystal cut parallel to the optic axis and examined in polarised light shows one extreme colour when the vibrations of the polarised light are perpendicular to the direction of the optic axis and the light is passing through the section as ordinary rays, and the other when the section is rotated in its own plane through 90° and the light is passing through the section as extraordinary rays. Uniaxial minerals are therefore often described as *dichroic*. Tourmaline shows the property of dichroism to an extreme degree. It crystallises in rhombohedral crystals of a long prismatic habit, often black, but less densely coloured specimens are also found. If a section is cut parallel to the long axis—which is also the optic axis—it appears green or brown in ordinary light. If the polariser is slipped into place and the section is oriented with its length parallel to the vibration direction of the rays, light traverses the section wholly as extraordinary rays and the section appears light brown or reddish. If now the stage is rotated, the colour may change to blue or green, but in all cases it darkens considerably, and when the vibrations are perpendicular to the long axis of the section and light is traversing it as ordinary rays, the section may appear almost black. In other words, the absorption of visible rays of all wave-lengths is far stronger for *o* than for *e*; indeed it is so strong for *o* that a slice about 3 mm. in thickness is entirely opaque to ordinary rays, while, unless the specimen is deeply coloured, the extraordinary rays are transmitted without very great loss of intensity.

As a result, tourmaline can be used to produce or to analyse polarised light, and jewellers use a small instrument known as the "tourmaline tongs" to examine the general refractive properties of a stone. It consists of two small plates of tourmaline cut parallel to the optic axis of the crystal and mounted parallel to each other at the extremities of a pair of tongs. The stone is held between the tongs; the vibration direction of light passing through each plate is indicated, and the relative orientation of the plates can be adjusted at will.

The two extreme colours shown by uniaxial minerals are known as *axial colours*; biaxial minerals in the same way show three different axial colours corresponding to the vibration directions giving the three chief refractive indices. The directions are sometimes denoted by the Gothic letters $\mathfrak{a}$, $\mathfrak{b}$ and $\mathfrak{c}$*; thus $\mathfrak{a}$ denotes the vibration direction of the fastest rays and one extreme colour of transmitted light, $\mathfrak{c}$ that of the slowest rays and another extreme colour, while

* It is becoming customary, especially in America, to substitute X, Y and Z for these directions so as to avoid the use of obsolete type and to prevent possible confusion with the A, B and C crystallographic axes.

𝕓 is the third perpendicular vibration direction and produces a third characteristic colour. In the description of
the pleochroism of a mineral the axial colour produced by
vibrations in each direction is quoted: e.g. for biotite, 𝕒 yellow
or green, 𝕓 and 𝕔 dark brown.

When a crystal is examined in ordinary non-polarised
light, the colour of any face will be a mixture of the colours
of the two rays vibrating parallel to that face. Hence the
combination of two axial colours produces what is known as
a *face colour*, and this will vary with variation in the axial
colours. For example, in a rock section containing biotite
sections cut parallel to the hexagonally shaped basal plane
will show a mixture of the 𝕓 and 𝕔 colours, which are both
dark brown, while lath-shaped sections perpendicular to
this plane may show a much lighter tint, due to the combination of the brown of 𝕓 or 𝕔 with the yellow tint of 𝕒.

Face colours are very strikingly shown in the case of the
rhombic mineral cordierite. Vibrations parallel to the
brachy-axis give as axial colour grey-blue; vibrations
parallel to the macro-axis give a deep blue, and the third
axial colour is yellow. Consequently the macro-pinacoid
(100) appears bluish-green, owing to the combination of
the yellow and the grey-blue, the brachy-pinacoid (010)
appears yellow-green, a mixture of the yellow and deep
blue, while the basal plane (001) appears blue.

C. *Examination between Crossed Nicols.*

Interference colours. If after the examination of a rock
slice with only the polariser in action the analyser is also
slid into place in the crossed position, a striking transforma-
tion takes place. Transparent minerals of low birefringence
form a patchwork in black, white and grey; others of higher

birefringence show pure, brilliant tints which contrast strongly with the duller tones, while minerals of unusually strong double refraction, such as calcite, exhibit less pure and more delicate colours, pinks, blues and greens. These are the interference colours whose method of production was described in an earlier section (see p. 114). An interference tint is produced by the mixture of residual colours left after certain wave-lengths have been removed from the original white light by the process of interference. Grey is produced by partial interference of all wave-lengths without the total extinction of any; yellow and orange result from the extinction of wave-lengths in the region of the blue and violet, blue from the extinction of the red and orange, green from a total or partial extinction of the blue and red simultaneously.

If the stage of the microscope is rotated, the interference colour of a mineral is seen to attain maximum intensity in four positions 90° apart; at four intermediate positions, known as *positions of extinction*, the mineral becomes dark. Interference tints should be studied when the mineral is oriented to show the maximum intensity of colour.

A more detailed investigation into the history of rays passing successively from the condenser through the polariser P, the crystal section C and the analyser A will make the phenomenon clearer. Fig. 111a shows in vertical section the path of the light rays, while fig. 111b shows the directions of vibration of the light at each stage. In fig. 111a the eye is looking horizontally at the vertical microscope tube; in fig. 111b it is looking vertically down on to each medium in turn, polariser, air-space, crystal, air-space, analyser, and air-space immediately above it.

The light leaves the polariser P vibrating along the shorter diagonal of the nicol, which is represented here as

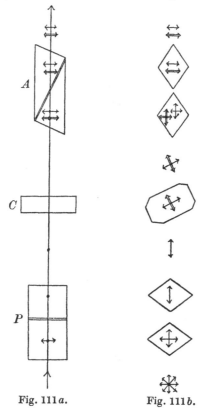

Fig. 111 a.   Fig. 111 b.

Passage of a ray through crystal section between crossed nicols.

being perpendicular to the plane of the paper. These vibrations strike the lower surface of the crystal section C, the rays being normal to the surface. Within the crystal the vibrations will be resolved into two equivalent sets of vibrations parallel to the vibration traces in that section; the rays produced may travel with different velocity and hence may be out of phase with one another by the time they pass out into the air-space above the crystal. They are therefore denoted by different types of line in fig. 111. They will continue as two separate rays until they enter the analyser, when each will be resolved into a pair of components vibrating parallel to the vibration traces of the calcite. Of each pair only that component will pass into the upper half of the nicol which is vibrating parallel to the plane of the paper. Hence two rays will issue from the top of the analyser, both derived from the same source and so having the same period, both vibrating in the plane of the paper, both travelling along the same path, but in the general case with a definite difference of phase. These are the necessary conditions for the production of interference; it remains to decide for what phase difference the two rays will annihilate each other and for what phase difference they will reinforce.

In fig. 112 let PP' be the vibration direction of the polariser, AA' that of the analyser, and XX', YY' the two vibration directions for the crystal section. When a vibration from the polariser, represented in intensity and phase by the line Op, enters the section, it will be resolved into two components, Ox along OX, and Oy along OY. These two vibrations are in phase with each other when first set up, but when the ray has traversed a thickness T of the crystal one may have lagged behind the other by a definite amount Δ. If the retardation is equal to a complete number of wave-lengths, the vibrations will again be in phase when they leave the crystal, and Ox and Oy will still represent the vibrations. If, however, there has been a retardation equal to an odd number of half wave-lengths, the vibrations on leaving the crystal will be represented by Ox and Oy' (fig. 113). Taking the first case, suppose that there is a retardation of n wave-lengths, where n is zero or any whole number. The vibrations represented by Ox and Oy will travel through the air-gap and will fall on the analyser, where each will be resolved into a component parallel to AA' which will pass through the analyser, and a component

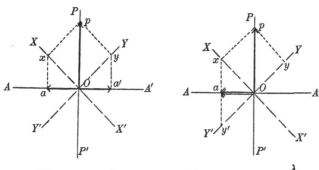

Fig. 112. $\Delta = n\lambda.$ Fig. 113. $\Delta = (2n+1)\dfrac{\lambda}{2}.$

Extinction of crystal section between crossed nicols.

in the perpendicular direction which will be totally reflected and therefore can be ignored. The vibration Ox will give a component Oa along AA', while Oy will give a component Oa'; it is obvious that these are equal but in opposite phase, and therefore that complete interference will occur, and no light will be visible through the analyser.

If, on the other hand, the phase difference produced by the crystal section is equal to $(2n+1)\dfrac{\lambda}{2}$, each of the vibrations Ox and Oy' will produce a component along AA' represented in phase and intensity by Oa; hence the two rays will reinforce one another, and the maximum amount of light possible will pass through the analyser.

If T is the thickness of the crystal slice, μ and μ' are the refractive indices of the two rays passing through the section, and v and v' their velocities, then $\mu = \dfrac{V}{v}$, and $\mu' = \dfrac{V}{v'}$, where V is the velocity of the rays in air. Moreover, the difference in the time taken by the rays to traverse the section will be $\dfrac{T}{v} - \dfrac{T}{v'}$, so that the retardation Δ will be equal to $V\left(\dfrac{T}{v} - \dfrac{T}{v'}\right)$, or, substituting for v and v',

$$\Delta = T\left(\mu - \mu'\right).$$

The condition for interference is therefore that

$$T\left(\mu - \mu'\right) = n\lambda,$$

and for reinforcement that

$$T\left(\mu - \mu'\right) = (2n + 1)\frac{\lambda}{2}.$$

If white light is used, the retardation may fulfil the condition for total interference for some value of λ, while for other values there may be partial interference, and for still other values there may be reinforcement. Hence the colour of the light seen through the analyser will be the residual tint produced after certain colours have been reduced or obliterated. The tint shown varies with the birefringence, measured by $\mu - \mu'$, and with the thickness T of the section. A rock section is usually cut to a thickness of from 0·03 mm. to 0·04 mm., and this produces for the birefringence of quartz, 0·009, a grey or yellowish tint. The thickness to which a section must be reduced for examination purposes is often decided by inspection of the interference colours exhibited by quartz or other well-known and easily recognisable minerals occurring in the section.

The variation of interference tint with thickness is well seen in the case of a long, narrow wedge cut from a colourless, uniaxial mineral, such as quartz, with the length of the wedge parallel to the optic axis. Light passing through the thickness of the wedge will be travelling perpendicular to the optic axis and will be doubly refracted. Consequently, if the wedge is examined between crossed nicols, the length of the wedge being at 45° to the vibration directions of the nicols, interference colours will be produced. The tint at any point will depend upon the thickness of the quartz at that point, and will therefore vary all along the length of the wedge. With monochromatic light of a definite wavelength the wedge transmits the colour peculiar to that wave-length, but is marked by a series of dark bands which cross it from side to side at the places where the thickness produces a retardation of λ, 2λ, 3λ, etc. If monochromatic light of a different wave-length is substituted, a similar effect is produced but the bands have shifted in position. If white light is used, the effect is virtually that of combining the colour and bandings produced by each wave-length separately, and at any point along the wedge the transmitted light will be lacking in any colour that is lessened or obliterated by interference at that point. Hence the wedge will show a series of tints which succeed each other in a definite order. The phenomenon was first described and explained by Newton, and the succession is known as Newton's Scale of Colours.

Fig. 114 shows how the scale is built up. Here we have representations of the quartz wedge as seen by light of six different wave-lengths, while below is listed the succession of colours produced by the combination of all six wave-lengths, which together may be taken as the equivalent

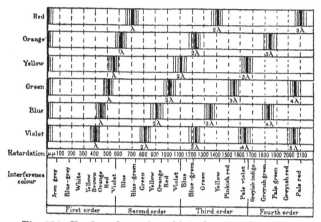

Fig. 114. To show formation of interference colour chart.

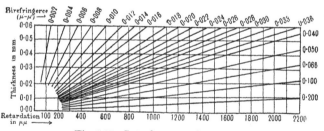

Fig. 115. Interference colour chart.

of white light, and also the retardation which produces each tint*. Thus, for a retardation of $300\mu\mu$ the intensity of the violet and blue rays is somewhat diminished, while yellow

---

* Wave-lengths and retardations of similar dimensions are usually expressed as multiples of a very small unit of length denoted by the symbol $\mu\mu$. $1\mu\mu$ is equal to one-millionth of a millimetre, or 10^{-7} cm.

and orange are at their maximum intensity. The result is white light mixed with an excess of yellow and orange, which produces a straw-yellow. At $550\mu\mu$ yellow is eliminated and all other colours are greatly reduced, with the exception of violet which is fairly strong. Hence the interference colour at this point is violet, but with a small change of retardation it alters quickly to red on the one hand and blue on the other. As the retardation increases, blue, green, yellow, orange and red appear in succession; at $1100\mu\mu$ violet appears again, for here the mid-spectrum colours are cut out, the violet is at a maximum and there is also a certain amount of red and blue which together accentuate the violet tint. This is followed by a repetition of the same series of colours; at $1650\mu\mu$ a third violet tint is produced, but now the colours are less pure, and for still greater retardations they become mixed with an increasing amount of white light and show softer, paler tones of pink and green. For convenience the succession is divided into *orders*; tints of the first order are produced when the retardation is less than 1λ for yellow light; second order colours are produced by a retardation greater than 1λ and less than 2λ, and so on for higher orders, the boundary of each order being marked by a violet tint. The first order colours are mainly greys and dull yellow and red; the second order colours are far more vivid, but in the third and fourth orders the intensity falls off until finally the separate tints are difficult to distinguish from each other.

A coloured chart of Newton's Scale of Colours is generally available for use in a petrological laboratory (fig. 115). It serves to show the connection between the interference tint and the strength of double refraction necessary to produce it in a section of definite thickness. The ordinates of the chart show values of the retardation Δ and the corresponding interference colours, the abscissae represent the thickness of the section, while radiating lines serve to connect the thickness and the retardation, each line corresponding to a definite value of birefringence. Thus, given a mineral section of a known thickness T, we can follow on the chart the horizontal line corresponding to T until we come to the colour shown by the section between crossed nicols. We shall then find ourselves on, or very near to, a radiating line marked at its extremity by the value for $\mu - \mu'$ that it represents, and we can accordingly read off the birefringence of the mineral under examination.

In any mineral the birefringence varies with the direction in which the section is cut, so that in a rock slice where the orientation of the minerals is haphazard the same mineral may exhibit different interference tints in different sections. Minerals of the cubic system show no birefringent effects, and between crossed nicols their sections remain dark. In uniaxial minerals the same is true of all sections cut perpendicular to the optic axis, since these also are singly refracting. The highest degree of birefringence will be shown by sections cut parallel to the optic axis; obliquely cut sections will show a lowered birefringence. Thus in a rock slice 0·04 mm. thick, sections of quartz will have the highest birefringence when cut parallel to the trigonal axis and will appear pale yellow; other obliquely cut sections will appear white or grey, while sections nearly perpendicular to the trigonal axis will show almost no birefringence. Tourmaline, which usually occurs in rhombohedral prisms, may show in the same rock slice triangular sections perpendicular to the trigonal axis which will be isotropic, and long lath-shaped sections, more or less parallel to the trigonal axis, which will show interference tints of the second order.

In biaxial crystals the birefringence will be at its maximum for sections cut parallel to the plane containing the α and γ vibration directions, that is to say, parallel to the optic axial plane, while sections accurately perpendicular to an optic axis will be singly refracting. Biotite is a monoclinic mineral with a tabular habit parallel to the basal plane (001), and sections parallel to this plane are hexagonal in outline. The optic axial plane is usually parallel to (010), and the Bx_a is nearly perpendicular to (001), so that in a rock slice we may find hexagonal sections which will exhibit birefringence of strength $\gamma - \beta$, and lath-shaped sections showing the characteristic basal cleavage, some of which exhibit the maximum birefringence $\gamma - \alpha$. This is considerably greater than $\gamma - \beta$, and the hexagonal sections therefore show decidedly lower interference tints than lath-shaped sections, though the latter may vary among themselves according as they are cut more nearly parallel to (010) or to (100). Variations in birefringence of this kind occur in the sections of all doubly refracting minerals, and in estimating the strength of birefringence that section must be selected which shows the highest interference tint.

Extinction. If a birefringent mineral section is rotated between crossed nicols, the intensity of the transmitted light is seen to vary, and in four positions 90° apart the light is entirely cut off. This effect, which is known as *extinction*, takes place when the vibration directions of the polariser and analyser are parallel to those of the crystal section; consequently the vibration directions of the section are sometimes called the *extinction directions*. Under these conditions there is no resolution of the polarised light on entering the section; it passes through with its plane of polarisation unchanged and is therefore entirely cut off by the analyser just as if the crystal section was not there.

The position of extinction depends therefore upon the vibration directions of a section. In uniaxial minerals for sections cut parallel to the optic axis, that is to say, for prismatic sections, extinction will occur when the vibration direction of one nicol is parallel to the prism edges which often form prominent outlines in such sections. In dome sections the extinction direction is parallel to the trace of one of the horizontal axes; in both cases the extinction directions are parallel to prominent crystallographic edges, and the sections are said to show *straight extinction* with reference to these edges. From fig. 92 it is easy to see that every section of a uniaxial crystal will show straight extinction with reference to a crystallographic axis or the trace of a crystallographic axial plane in that section. Rhombohedral sections will show straight extinction with reference to the diagonals of the rhombohedron. In rhombic crystals the three principal vibration directions are parallel to the crystallographic axis, and hence sections of rhombic crystals also show straight extinction.

Sections of monoclinic and triclinic crystals usually show *oblique extinction*, that is to say, the extinction or vibration directions in a section are not parallel to crystallographic edges. The angle between the extinction direction and a prominent crystallographic edge, usually one parallel to the vertical axis, is known as the *extinction angle* of the section with reference to that edge. The size of the extinction angle is a characteristic of the mineral and is a useful guide to its identity, especially in the case of the various members of a mineral group. For example, in the case of the triclinic plagioclase felspars, the progressive variation in chemical composition from albite to anorthite is accompanied by a progressive change in the value of the extinction angle. These minerals possess cleavages parallel to (001) and to (010); the cleavage traces can therefore be used as a guide to the identification in a section of the edge between these two faces, and the extinction of the plagioclases is usually quoted with reference to this edge. For sections parallel to (001) the extinction angle varies from 5° on one side of the reference line for albite to 37° on the other side of the line for anorthite, and for sections parallel

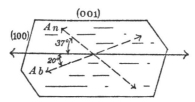

Fig. 116. Extinction directions in plagioclase felspars.

to (010) there is a similar variation from 20° on one side of the line for albite to 37° on the other side for anorthite (fig. 116).

Sections of monoclinic crystals parallel to the *B* axis show straight extinction, since this axis is always a principal vibration direction. Thus sections of the monoclinic felspar, orthoclase, cut parallel to (001), to (100), or more generally, to $(h0k)$, show straight extinction with reference to the *B* axis, and this helps to distinguish it in sections from the triclinic plagioclase felspars.

The extinction angle is measured by setting the crystal in the extinction position. A reading is taken on the stage scale, and the stage is rotated until the reference line, crystal edge or cleavage trace, is parallel to one of the cross-wires of the microscope which themselves are parallel to the vibration directions of the nicols. A second reading is taken on the stage scale, and the angle of rotation thus measured is the angle of extinction. In some minerals there is so great a difference in the extinction angle for light of different wave-lengths that with white light complete extinction can never be obtained, and for accurate determinations monochromatic light has to be used.

A convenient device for giving greater accuracy to the setting for the extinction position is the *gypsum plate*. This is cut from a crystal of gypsum parallel to the optic axial plane, and is of such a thickness that between crossed nicols it gives a reddish-violet of the first order; any slight alteration of the retardation between the rays alters the tint very quickly to a blue of the second order on the one hand and a first order yellow on the other, so that this colour is often referred to as the *sensitive tint*. There is usually an arrangement whereby a gypsum plate can be slid into the tube of the microscope just below the analyser, oriented with its vibration directions at 45° to those of the nicols. A doubly refracting section whose vibration directions are accurately parallel to those of the nicols will have no effect on the interference tint shown by the gypsum plate, but a very slight deviation from parallelism will introduce an additional amount of retardation and will at once alter the colour, so that it is easy to adjust the crystal section to its exact position of extinction.

The phenomenon of extinction reveals regions of different structure or composition in a crystal which may be invisible under ordinary conditions. For example, the two halves of a twin crystal often extinguish in different positions, and, if the twin-plane is also the composition plane, the extinction positions will be symmetrical about the line of junction. Or again it is not unusual for a mineral such as plagioclase to possess a zoned structure, the successive zones being of different chemical composition. In this case the zones may have different positions of extinction, and will thereby show up more clearly than in ordinary light.

Directions of Slow and Fast Vibration. The position of extinction gives the two directions of vibration in a bire-fringent section, and one set of vibrations travels more quickly than the other. To determine which is the direction of the faster ray and which of the slower, various devices are used, the commonest being the *quartz wedge* and the *quarter-wave-length mica plate.*

The position of extinction is found and the section is turned through 45° so that its interference colour is at its maximum brightness. The quartz wedge is slid into the microscope just below the analyser, also in the 45° position, so that its vibration directions are parallel to those of the section. Quartz being optically positive, the vibrations parallel to the length of the wedge travel more slowly than those across the wedge. If the crystal section is oriented so that its vibration direction for slow rays is parallel to the length of the quartz wedge, then, as the wedge is slowly pushed across the field of view, the retardation between the two rays in the section is still further increased in passing through the quartz, and the interference colour rises con-tinually as a thicker and thicker section of the wedge comes into the path of the rays. If, however, the fast vibration direction in the crystal section is parallel to the length of the wedge, that ray which travels faster in the crystal becomes the slower ray in the quartz, and the wedge will act as a compensator. If the wedge is pushed slowly in, the colour of the section falls until finally the thickness of the wedge is such as to produce equal and opposite retardation to that of the section, and a dark band will appear across the field of view.

A simpler device for the same purpose is the *mica plate*, a cleavage flake of white mica of such thickness that it produces a retardation of a quarter of a wave-length for yellow light; it is accordingly known as the *quarter-wave plate*. Between crossed nicols it gives a pale blue-grey of the first order; the plate is marked with a double-headed arrow which indicates the vibration direction of the slow ray. Usually the petrological microscope is provided with a slot below the analyser into which the mica plate can be slid. As before, the crystal section is placed in the 45° position and the mica plate, also in the 45° position, is slid into the field of view. A rise in interference tint due to the mica plate indicates that the slow vibration direction in the section is parallel to the slow direction of the mica plate; a fall in colour indicates that the slow direction in the section is perpendicular to that in the mica plate.

The sensitive plate of gypsum can be used for the same purpose.

Crystals of prismatic habit usually have one vibration direction parallel to the long axis, and the mica plate affords a ready means of showing whether this axis is the vibration direction of the slow or the fast ray. If the slow ray vibrates parallel to the length of the crystal section the crystal is said to have *positive elongation*; if the fast ray has this vibration direction the crystal is said to have *negative elongation*. In the very common case of uniaxial crystals elongated along the unique crystallographic axis, which is also the optic axis, this convention implies that an optically positive crystal has positive elongation.

D. *Examination in Convergent Polarised Light.*

Since the interference tint shown by a crystal between crossed nicols depends upon the orientation of the crystallographic axes with regard to the direction of the ray and also upon the thickness of the crystal traversed by the ray, a beam of converging rays will not produce a uniform colour when passed through a crystal section between crossed nicols.

Suppose that a convergent beam of polarised rays passes through a section of a uniaxial crystal cut perpendicular to the optic axis, so that the central ray of the beam is normal to the section and travels through the crystal parallel to the optic axis. All other rays are oblique, and as their inclination to the axis increases there is a corresponding increase in birefringence along their direction and in the length of path that they traverse within the crystal. For both reasons the retardation between the *o* and *e* rays travelling along the same path will increase as their path becomes more oblique to the central ray of the beam. At a certain inclination the pair of rays will leave the crystal with a phase difference of λ for light of some particular wave-length, and consequently they will interfere for light of this wave-length. All rays having this definite inclination will lie on the surface of a cone described about the optic axis (fig. 117), and they will be brought to a focus by the objective in a circle about a central point which is the focus of rays normal to the section (fig. 118). In the same way, other pairs of rays at a greater inclination will have a phase difference of 2λ, and will form a second cone of wider angle about the optic axis; these will be brought to a focus in a larger circle concentric with the first. Fig. 117 shows a crystal section with the cones of rays having phase differences of λ, 2λ and 3λ. Examined under these conditions in white light, therefore, a section of a uniaxial crystal cut perpendicular to the optic axis shows a succession of concentric coloured rings of the same nature as the inter-

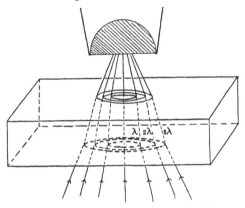

Fig. 117. Crystal section in convergent beam.

ference tints discussed in the last section. The succession of colours from the centre outwards is that of Newton's Scale, and the central spot, representing the direction of the optic axis, will be dark.

The interference picture (fig. 118) is focussed by a high-power objective and the image is out of range of the eye-piece; by removing the eye-piece it can be seen by the naked eye, small and sharp in outline, or it can be brought within the range of the eye-piece by inserting an extra lens, the Bertrand lens, just above the analyser. It is seen to

better advantage if monochromatic light is used; then the
dark rings marking retardations of λ, 2λ, 3λ ... stand out
plainly against the coloured background. The rings are
seen to be crossed by two dark bands lying parallel to the
vibration directions of the nicols; the cause of these becomes clear on considering the vibration directions of the
various rays making up the picture (fig. 119).

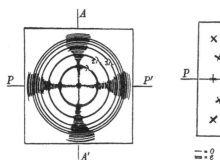

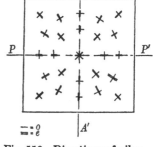

Fig. 118. Interference figure   Fig. 119. Directions of vibra-
for uniaxial crystal.   tion of convergent rays.

Every ray traversing the section obliquely to the optic
axis is divided into an ordinary ray vibrating perpendicular
to the optic axis, that is to say, tangentially to the circles
of the optic picture, and an extraordinary ray vibrating in
the plane of the optic axis, that is to say, in a radial direction in the optic picture. The little crosses in fig. 119 indicate the vibration directions at each point of the optical
picture, the o and the e vibrations being indicated by
different types of line. Along the vibration directions of the
nicols, indicated by PP' and AA', these vibrations are
parallel to those of the nicols, so that for rays emerging in
points along PP' or AA' the crystal is in the extinction

position, and the lines appear dark. If the section is rotated or shifted from side to side on the stage the interference figure remains stationary, and the positions of the so-called rings and brushes remain unaltered; this emphasises the fact that the optic axis of a crystal is a direction and not a specific line. The only way to shift the interference figure is by tilting the section; the slightest change in the direction of the optic axis causes a shift of the whole

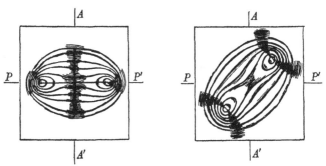

Fig. 120. Parallel position.   Fig. 121. 45° position.
Interference figures for biaxial crystal.

figure. For this reason the figure will only appear central and symmetrical if the section is truly normal to the optic axis; sections slightly inclined to the axis will appear asymmetric, and in this case the centre of the figure will rotate with the rotation of the section.

The production of interference figures is useful in distinguishing between a uniaxial and a biaxial crystal, for the latter shows a different arrangement of rings and brushes. To obtain the interference figure a biaxial crystal must be cut perpendicular to the acute bisectrix; the optic axes are then symmetrically oblique to the normal to the section and in the interference figure their points of emergence are indicated by two "eyes", each surrounded by oval curves, outside which are oval curves laterally depressed, embracing both centres. When the line joining the eyes is parallel to the vibration direction of either nicol, this direction is marked, as in the uniaxial figure, by a dark line running across the rings, while a second dark band runs perpendicularly between the eyes (fig. 120). If, however, the section is rotated to the 45° position, the figure undergoes a change. The eyes and rings remain unchanged, but the two dark lines part in the centre and rejoin to form hyperbolas passing through the eyes and curving outward from the centre (fig. 121). This change affords an unmistakable distinction between sections of uniaxial and biaxial crystals, even in cases when, owing to oblique cutting, only one eye of the biaxial figure is visible, giving the figure a superficial resemblance to that of a uniaxial crystal. With white light the interference figure may show striking colour effects; dispersion of the optic axes and of the acute bisectrix may be visible, and the degree of crystallographic symmetry may be apparent in the degree of symmetry shown by the disposition of colour (see p. 142).

Interference figures can be used also to indicate the optical sign of a crystal.

A section of a uniaxial crystal is adjusted to show the interference figure; a quarter-wave-length mica plate is then inserted in its slot in the 45° position. The figure shows a change: the dark cross disappears, since the mica causes resolution of the vibrations formerly parallel to PP' and AA' (fig. 122); the rings break up into quadrants, those in two opposite quadrants closing in while in the alternate quadrants they expand, and two black patches appear near the centre.

Suppose that the slower vibration direction of the mica plate is parallel to the line bisecting quadrants 1 and 3. If in these two quadrants the rings close in, the birefringent effect of the mica is obviously increasing that of the crystal along this line, since a smaller length of path through the crystal is now required to produce a total retardation of λ.

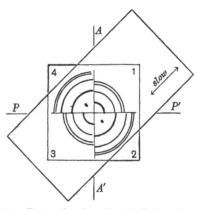

Fig. 122. Determination of optical sign of uniaxial crystal by use of mica plate.

Hence the slow vibration direction in the crystal plate must be parallel to that in the mica plate; that is to say, referring to fig. 119, the extraordinary rays travel more slowly than the ordinary rays, and the crystal is therefore optically positive. The black dots in quadrants 2 and 4 represent places where the negative retardation of the mica exactly counterbalances the positive retardation in the crystal; the area between these and the centre is light by reason of the under-compensated birefringence of the mica in this area. It may aid the memory to look upon the two black spots as forming with the slow direction of the mica plate the + sign of the positive crystal.

A similar effect is produced when the quarter-wave plate is used with a biaxial section. If the section is oriented so that the eyes lie along the vibration direction of one of the nicols, the positive crystal is indicated by a contraction of the rings in quadrants 1 and 3 and their expansion in quadrants 2 and 4, just as in the case of a uniaxial crystal.

Another method of determining the sign of a biaxial crystal is by sliding over the interference picture in the 45° position the thin quartz wedge described earlier. If the length of the wedge is parallel to the axial plane of the crystal—that is to say, parallel to the line of the eyes, the rings will be seen to open or close, the amount of movement increasing with the thickness of the wedge brought into play. If the wedge causes expansion of the spaces between the rings, the quartz must be detracting from the birefringent effect of the crystal; this means that the length of the quartz, which is the direction for the slower vibrations in the quartz itself, must be the direction of the faster vibrations in the crystal section. This direction is parallel to the obtuse bisectrix in the crystal, so that the acute bisectrix must be the direction of the slowest rays, and the crystal is therefore optically positive. The result should be checked by pushing in the wedge with its length perpendicular to the line of the eyes: if the crystal is positive the rings will now appear to move in towards the optic axes. Both effects are reversed in the case of a negative crystal.

For the use of interference figures in measuring the dispersion of the optic axes and bisectrices and for finding the value of the optic axial angle, reference must be made to the standard text-books.

BIBLIOGRAPHY

General:

T. V. Barker. *The Study of Crystals.* London, 1930.

E. S. Dana (revised W. E. Ford). *Text-book of Mineralogy.* (4th ed.) New York and London, 1932.

H. A. Miers. *Mineralogy.* (2nd ed., revised H. L. Bowman.) London, 1929.

A. F. Rogers. *Introduction to the Study of Minerals and Rocks.* (2nd ed.) New York, 1921.

A. E. H. Tutton. *Crystallography and Practical Crystal Measurement.* London (1st ed.), 1911, and (2nd ed.), 1922.

Nature and Classification of Crystals:

D. B. Briggs. *The Study of Crystals.* London, 1930. [Elementary.]

W. J. Lewis. *A Treatise on Crystallography.* Cambridge, 1899.

Internal Structure:

W. H. and W. L. Bragg. *X-Rays and Crystal Structure.* (4th ed.) London, 1924.

W. H. Bragg. *An Introduction to Crystal Analysis.* London, 1928.

W. A. Caspari. *The Structure and Properties of Matter.* London, 1928.

R. W. James. *X-Ray Crystallography.* London, 1930.

Optical Properties:

J. K. Robertson. *Introduction to Physical Optics.* London, 1929.

R. W. Wood. *Physical Optics.* (3rd ed.) New York and London, 1928.

N. H. and A. N. Winchell. *Elements of Optical Mineralogy.* Part I. (4th ed.) New York and London, 1931.

Methods of Investigation:

J. W. Evans. *The Determination of Minerals under the Microscope.* London, 1928.

A. Johannsen. *Manual of Petrographic Methods.* (2nd ed.) New York and London, 1918.

H. G. Smith. *Minerals and the Microscope.* (2nd ed.) London, 1927.

E. Weinschenk (transl. by R. W. Clark). *Petrographic Methods.* New York, 1912.

N. H. and A. N. Winchell. *Elements of Optical Mineralogy.* Part I. (4th ed.) New York and London, 1931.

INDEX

TABLE OF CRYSTAL SYSTEMS

System	Crystallographic Constants	Essential Symmetry	Number of Classes
Cubic or Regular	Three equal, rectangular axes $a:b:c = 1:1:1$ $\alpha = \beta = \gamma = 90°$	4 triad axes 3 diad *or* 3 tetrad axes, coinciding with crystal axes	5
Tetragonal	Three rectangular axes, two equal $a:b:c = 1:1:?$ $\alpha = \beta = \gamma = 90°$	1 tetrad axis, simple or alternating, coinciding with C axis	7
Orthorhombic or Rhombic	Three rectangular axes, unequal $a:b:c = ?:1:?$ $\alpha = \beta = \gamma = 90°$	3 diad axes, coinciding with crystal axes *or* 1 diad axis and 2 perpendicular planes, intersecting in diad axis	3
Monoclinic or Oblique	Three axes, one perpendicular to the other two, which are not at 90° $a:b:c = ?:1:?$ $\alpha = \gamma = 90°;\ \beta = ?$	1 diad axis, coinciding with B axis *or* 1 plane perpendicular to B axis	3
Triclinic or Anorthic	Three axes, all inclined $a:b:c = ?:1:?$ $\alpha = ?,\ \beta = ?,\ \gamma = ?$	No axes or planes	2
Hexagonal	Three equal coplanar axes at 60°, a fourth unique axis perpendicular to the other three $a_1:a_2:a_3:c = 1:1:1:?$	1 hexad axis, coinciding with unique crystal axis	5
Rhombohedral or Trigonal	*Either* as in hexagonal system, *or* three equal axes, equally inclined but not at 90° $a:b:c = 1:1:1$ $\alpha = \beta = \gamma = ?$	1 triad axis, parallel to unique crystal axis of hexagonal system *or* 1 triad axis, equally inclined to three equal crystal axes	7

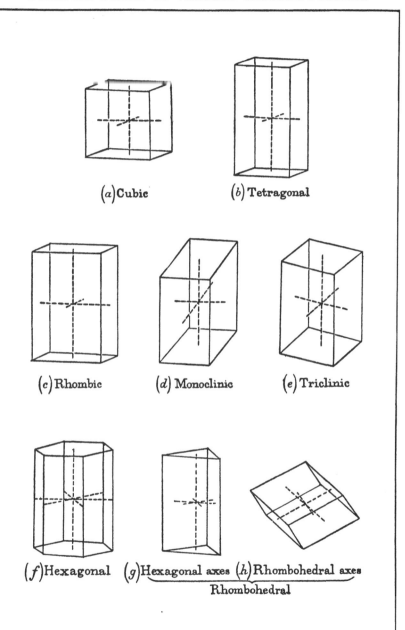

(a) Cubic (b) Tetragonal

(c) Rhombic (d) Monoclinic (e) Triclinic

(f) Hexagonal (g) Hexagonal axes (h) Rhombohedral axes

Rhombohedral